世界经典建筑设计丛书

教育机构

韩国 A&C 出版社　著

北京科学技术出版社

图书在版编目（CIP）数据

教育机构 / 韩国A&C出版社著 ; 北京科学技术出版社译. -- 北京 : 北京科学技术出版社, 2019.1

（世界经典建筑设计丛书）

书名原文：Educational Services

ISBN 978-7-5304-9393-9

Ⅰ. ①教… Ⅱ. ①韩… ②北… Ⅲ. ①教育建筑—建筑设计—作品集—世界—现代 Ⅳ. ① TU244

中国版本图书馆 CIP 数据核字（2018）第 063374 号

世界经典建筑设计丛书 · 教育机构

作　　者：韩国 A&C 出版社
策划编辑：陈　伟
责任编辑：王　晖
封面设计：芒　果
责任印制：张　良
出 版 人：曾庆宇
出版发行：北京科学技术出版社
社　　址：北京西直门南大街 16 号
邮政编码：100035
电话传真：0086-10-66135495（总编室）　0086-10-66113227（发行部）
　　　　　0086-10-66161952（发行部传真）
网　　址：www.bkydw.cn
电子信箱：bjkj@bjkjpress.com
经　　销：新华书店
印　　刷：北京捷迅佳彩印刷有限公司
开　　本：920mm×1260mm 1/32
字　　数：237 千字
印　　张：8.5
版　　次：2019 年 1 月第 1 版
印　　次：2019 年 1 月第 1 次印刷
ISBN 978-7-5304-9393-9/T・970

定　　价：138.00 元

>>CONTENTS

教育机构

>> EDUCATION

EWHA CAMPUS COMPLEX

Location Daehyeon-dong, Seodaemun-gu, Seoul, Korea **Site Area** 19,000m^2 **Building Area** 70,000m^2 **Building Volume** 350,000m^2 **Landscape Area** 31,000m^2 **Start of Conceptual Design** 2004 **Construction Start** 2005 **Completion** 2008.4 **Architecture Design** Dominique Perrault Architecture(Paris) **Partner** BAUM Architects, Engineers & Consultants, Inc.(Seoul) **Design Team** DPA_Gaelle Lauriot Prevost, Simon Guillemoz / BAUM_Won Hyung-joon, Son Sung-ho, Shin Jeong-hoon **Engineer** Perrault Projets(Architectural Engineering, Paris); VP&Green Ingenierie(Structural Engineers, Paris); HL-PP Consult(Building Services, Munich), Jean-Paul Lamoureux(Acoustic); Rache-Willms(Facades, Aachen) **Consultant** Jeon and Lee Partners(Structural Engineer, Seoul), HIMEC(Mechanical Engineer, Seoul), CG E&C(Civil Engineer, Seoul), CnKb Associates(Landscape, Seoul) **Client** Ewha Campus Center Project T/F, Ewha Womans University **Photographer** Namgoong Sun, Dominique Perrault Architecture

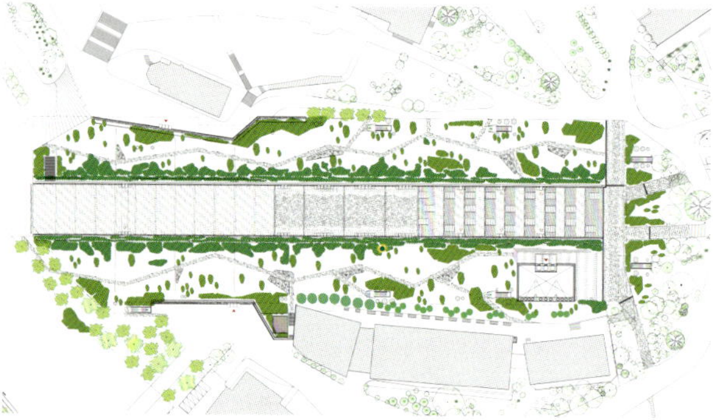

SITE & ROOF PLAN

The new campus center of Ewha Womans University was inaugurated in April 2008 and accommodate 20,000 students. A contemporary metropolis as Seoul seems to have a dramatic need of building who creates a breath in the city to balance the fast densification of real estate developers. By burying the complex, I have the desire to appropriate the territory, to mingle the constructed material with the ground, the desire to exploit to its paroxysm the idea that 'concept and matter have to grapple one with another'. The tree sunken architectures buildings are buried to liberate the territory above. The genealogy with the previous projects, as in the French National Library in Paris with its sunken garden in the middle of a wood deck framing, or in the Velodrome and Olympic swimming pool in Berlin engraved in an apple tree orchard, is evident when we consider the fantastic excavation required by this construction.

A SUSTAINABLE CAMPUS The idea of this underground university was born out of the desire to preserve a large green space in the center of the university campus. The six-stories building is organized around a long ramp sloping down in opposition with the gently rising natural topography. The two large glass curtain walls facing the external circulation ramp serve as elements for day lighting and allow natural ventilation of parts of the building as well. In fact, the first architectural ideas that give its strong identity to the building enable extraordinary performances in terms of stainability.

THE HIDDEN VALLEY OF EWHA First, the new entrance leads to a transversal platform. A landscape then, more than an architectural work, 'The Campus Valley' is engraved into the ground and marks the entrance of University site like a furrow. Starting by a long slopping down esplanade, it rise up at his edge with benches and steps which can be used as an outdoor amphitheatre. The two mirror facades erected face to face design two high glass cliffs who bring natural light to the buried spaces mixing opportunely the study places, administration with cultural and commercial. Designed and rigorously built to resist to high wind stress, with its very material, strong and heavy the full stainless steel blades, bolts, brackets and canopies are disappearing due to the huge scale. In the middle of the valley, only remains become dreamlike immersion between two cascades of light. All together, by using simple shapes which allow the materials to show their pure appearance, we can reach an immaterial effect which participates to the disappearance of the building. Finally the building disappears under the rooftop garden. It has a very natural aspect, walk paths are made of rough natural stones, evergreen bushes and flower trees like spiraea prunofolia, rhododendron, kerria, are planted randomly. This harmony reinforces the disappearance of the building.

PROCESS Due to the particular position of its stainless steel fin, the realised curtain wall is the result of a long study involving Korean, German, French engineers and architects. In order to resist to the wind load, the aluminium mullion stiffness of the 17m high curtain wall needs to be reinforced with a steel wale in order to resist to the particularly difficult wind conditions, typhoons of Korea. Normally, the wale is placed inside the building, just behind the aluminium vertical profile. By placing it outside, by shaping it as a fin and by using a mirror polished finish we transform those structural elements into a visual experience and offer to the building and users an installation similar to a contemporary art work.

During construction design phase, we designed it as a completely flat plate in order to get a pure geometry. By connecting the 5.4m invisible fitting screws hidden in the 20mm thickness of the stainless steel plate, we wanted to give the impression that they are made in one piece. The fact that the fin is outside complicates the work : the wind going laterally into the Valley has an effect on it. In order to resist to the wind stress, we added horizontal brackets perpendicular to the fin. Secondly, we have to make the fin connected to the mullion in order that the last one benefits the stiffness of the first one. When both elements are inside, the connection is easier, but as the wale is outside, we have a polyamide insulating element in between. As this element has no structural qualities, we had to connect punctually the mullion and the fin. This connexion creates a very local thermal bridge. During the mock ups testing phase, we realised that the theoretical exactitude required during construction design phase was too difficult to reach. The workmanship quality was too difficult to maintain in industrial production conditions. We had to replace the fitting screws by cap nuts bolts with treated rods to connect the fins together. The result is amazingly more interesting than the previous version. At a small scale we have a much more material, heavy elevation. We even reinforced this aspect with full stainless steel canopies fixed to the fin by angles and similar bolts. But at a larger scale, because of the huge dimensions of the valley the connections seem to be a very little motive, dispatched randomly to animate the rigorous design. They don't get in conflict with the very immaterial visual perception, the impression of light reflection. Text by Dominique Perrault Architecture

1 CINEMA
2 BANK
3 WELFARE FACILITIES
4 REST ROOM
5 MULTI-PURPOSE ROOM
6 THEATER
7 CEREMONIAL HALL
8 FITNESS CENTER
9 PLAZA
10 CAFE LOUNGE
11 PROFESSOR'S OFFICE
12 OFFICE
13 LECTURE ROOM
14 LANGUAGE LABORATORY
15 LECTURE HALL
16 GLOBAL LOUNGE

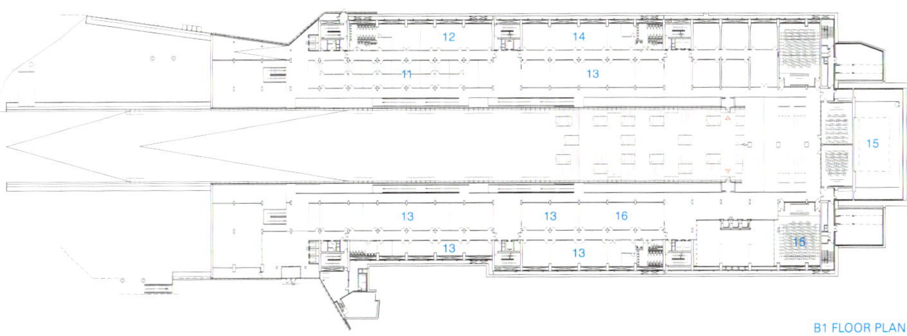

B1 FLOOR PLAN

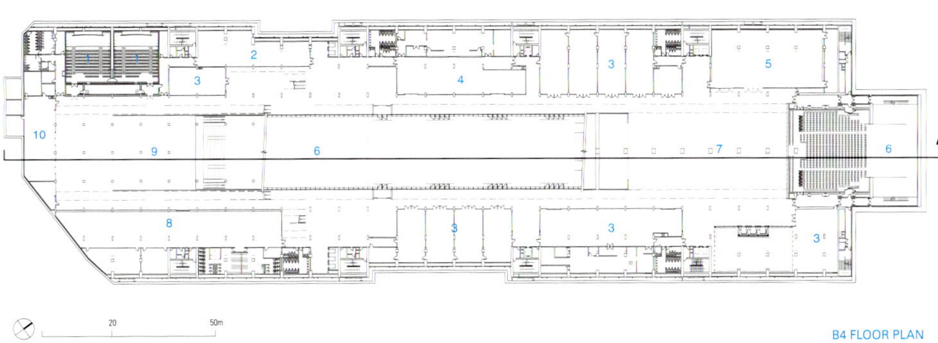

B4 FLOOR PLAN

The valley in the light seen from the main entrance to campus

In the valley

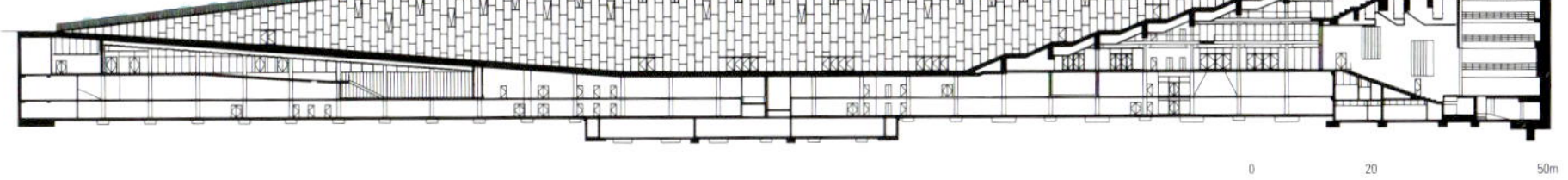

SECTION A

People inside can see through outside

Curtain wall detail

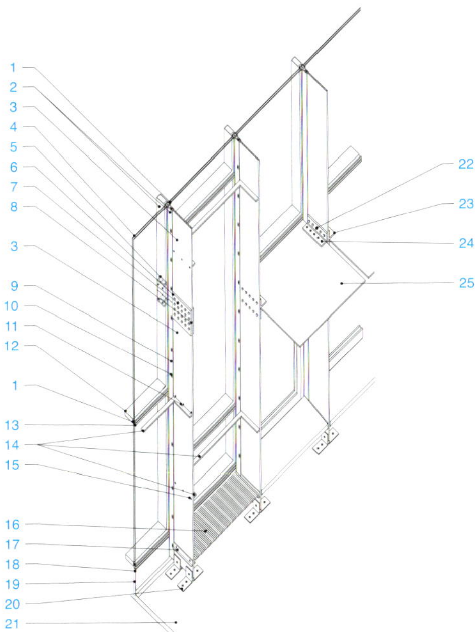

FACADE SECTION

1 DOW CORNING 993 STRUCTURAL GLAZING SEALANT OR SIMILIAR - CW PACKAGE
2 ALUMINUM PROFILE
3 POLISHED STAINLESS STEEL FIN
4 INSULATED GLASS UNIT TEMPERED GLASS GROUND FLOOR, HEAT STRENGTHENED GLASS UPPER FLOORS
5 SPLICED ALUMINUM PROFILE BOLTED TO INTERIOR STEEL REINFORCEMENT AS REQUIRED
6 POLISHED STAINLESS FULL THREADED ROD AND CAP NUT
7 POLISHED STAINLESS STEEL PLATE
8 SPLICE IN FIN WITH FIN LENGTH NOT TO EXCEED 5,400MM
9 SLOTTED MOLE(SLOT IN FIN) FOR VERTICAL DIFFERENTIAL MOVEMENT BETWEEN ALUMINUM AND STAINLESS STEEL
10 FLUSH THREADED STAINLESS STEEL BOLT EVERY 400MM
11 BRACKET BOLTED TO FIN FLUSH THREADED STAINLESS STEEL BOLT
12 HORIZONTAL ALUMINUM PROFILE
13 ALUMINUM COVER PLATE
14 POLISHED STAINLESS STEEL HORIZONTAL BRACKET
15 FLUSH THREADED STAINLESS STEEL BOLT FIN / BRACKET CONNECTIION
16 GALVANISED STEEL GRILL
17 ALUMINUM PROFILE - GRILL SUPPORT
18 ALUMINUM FLASHING
19 INSULATION
20 GALVANISED STEEL ANGLE SUPPORT
21 CONCRETE SLAB
22 POLISHED STAINLESS FULL THREADED ROD AND CAP NUT FIN / ANGLE CONNECTION
23 POLISHED STAINLESS STEEL ANGLE
24 FLUSH THREADED STAINLESS STEEL BOLT CANOPY / ANGLE CONNECTION
25 POLISHED STAINLESS CANOPY

The mirror facade reflects scenes of the exterior

FACADE HORIZONTAL SECTION B

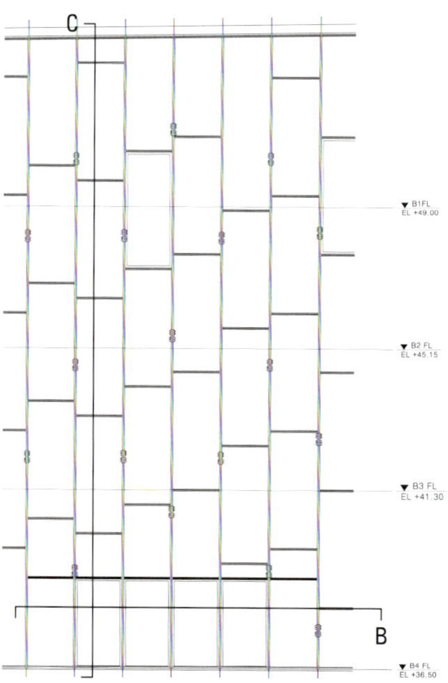

FACADE PARTIAL ELEVATION

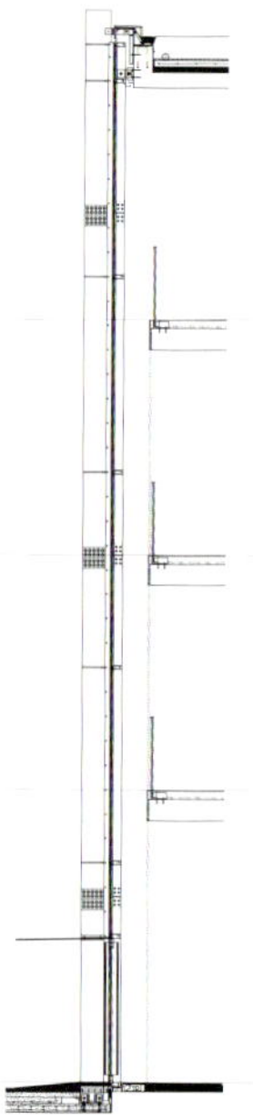

FACADE VERTICAL SECTION C

1 STAINLESS STEEL ANGLE MIRROR FINISH
2 STAINLESS STEEL PLATE T16 MIRROR FINISH
3 S'STL FLUSH THREADED BOLT
4 S'STL FULL THREADED ROD & CAP NUT M24X95
5 POLISHED STAINLESS STEEL BRACING FIN 20MM
6 INTERIOR
7 EXTERIOR
8 STAINLESS STEEL ANGLE SUPPORT T10 HAIRLINE FINISH
9 STAINLESS STEEL FLAT T10 HAIRLINE FINISH
10 FIN LENGTH 300MM
11 FIN LENGTH 450MM
12 FOLDED ALUMINUM PANELS
13 INSULATION
14 STEEL TUBULAR STRUCTURE
15 HEATING UNIT TO BE DEFINED OUTSWINGING DOOR-ALUMINUM
16 FRAME WITH INSULATED GLASS UNIT
17 VESTIBULE TO BE INCLUDED IN CURTAIN WALL PACKAGE
18 VERTICAL ALUMINUM PROFILE WITH STEEL REINFORCEMENT AS REQUIRED
19 HORIZONTAL ALUMINUM PROFILE
20 STEEL ANGLE
21 STAINLESS STEEL ANGLE
22 INSULATED GLASS UNIT
23 MIRROR POLISHED STAINLESS STEEL BRACING FIN 20MM
24 ALUMINUM SHEET
25 FULL HEIGHT STAINLESS STEEL TUBULAR HANDLE
26 GALVANISED STEEL GRILL BY MEISER GRUPPE OR SIMILIAR
27 ALUMINUM ANGLE - GRILL SUPPORT
28 THERMAL INSULATION - CW PACKAGE
29 FLASHING - CW PACKAGE
30 BRICK
31 STAINLESS STEEL FLAT T16 POLISHED FINISH
32 WATERPROOF MEMBRANE
33 GLAVANISED STEEL ANGLE - CW PACKAGE
34 CONCRETE BASE FOR FACADE SUPPORT BEYOND
35 TUFTI GUARD RUBBER STRIPS IN ALUMINUM GRID FRAME

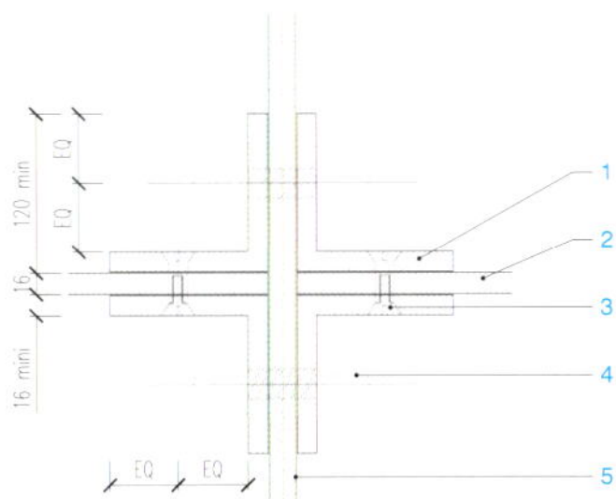

FACADE JOINT SECTION D

The main stairs of the valley permeate inner space naturally

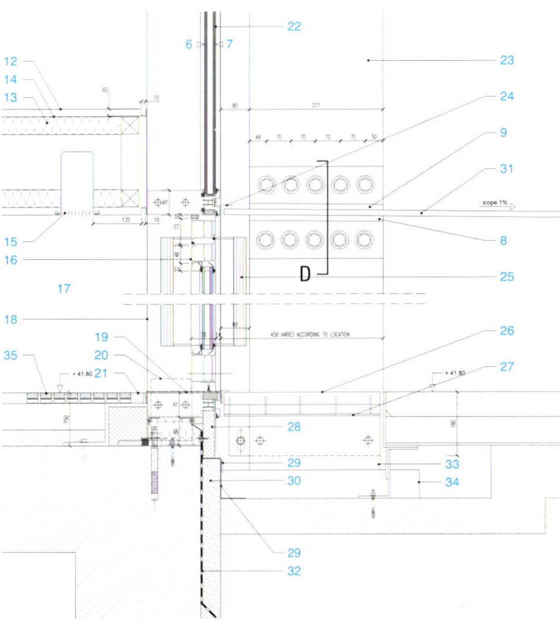

FACADE CANOPY SECTION

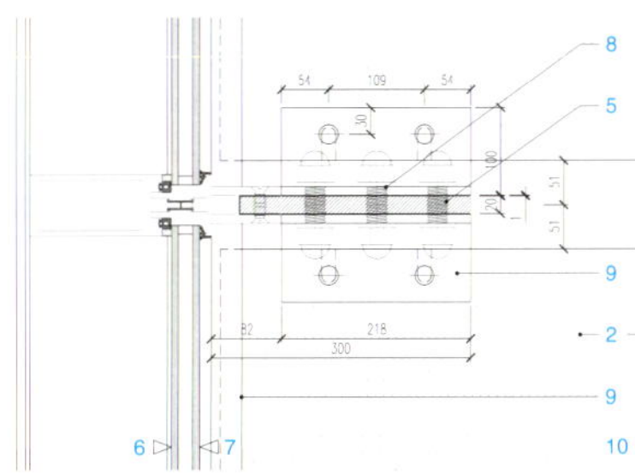

JOINTTYPE 300MM DETAIL

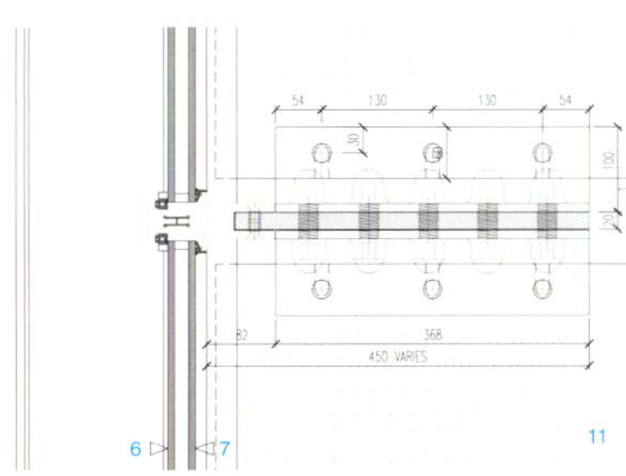

JOINTTYPE 450MM DETAIL

A corridor links with every floors and rooms

Because of hidden mass, when people enter the inside, they feel the largest space sense about huge scale

Ceremony hall under the main stairs of the building

Buildings are buried underground

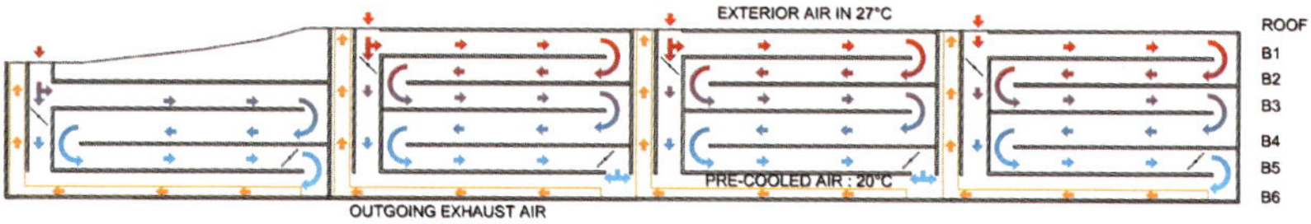

THERMAL LABYRINTH(SUMMER)

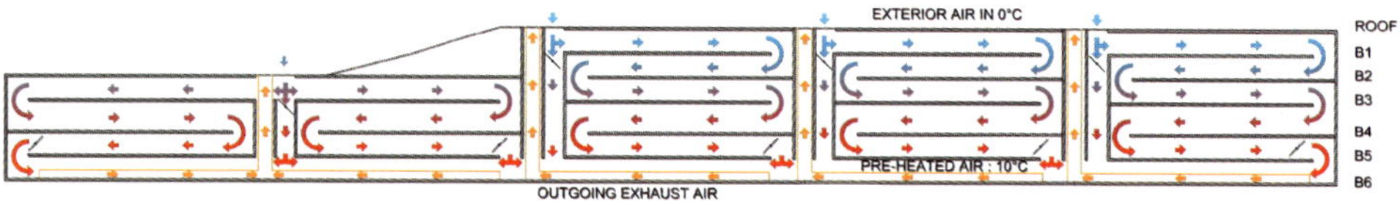

THERMAL LABYRINTH(WINTER)

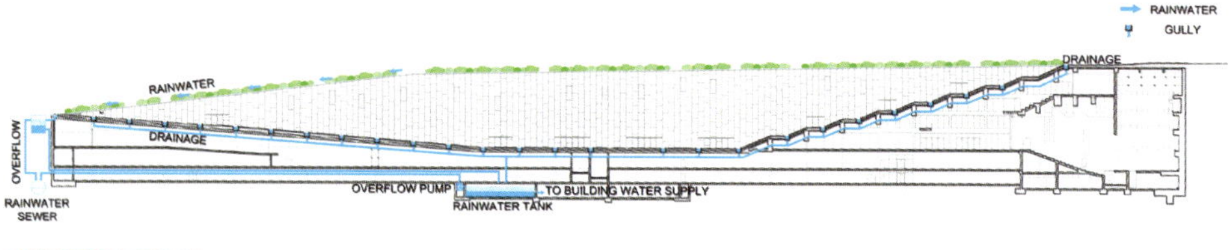

RAINWATER REUSE PLAN

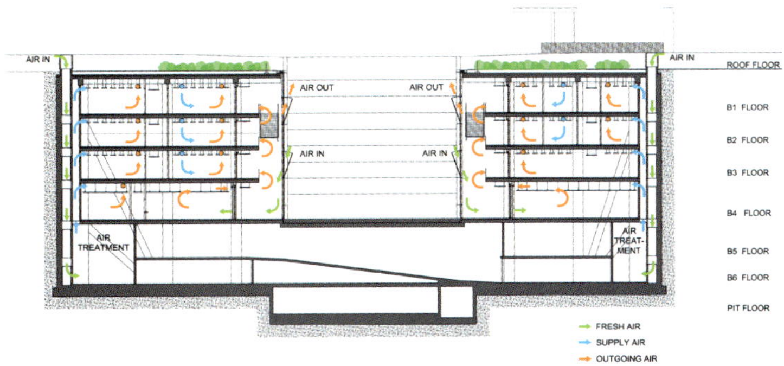

AIR CIRCULATION PLAN(SPRING & AUTUMN)

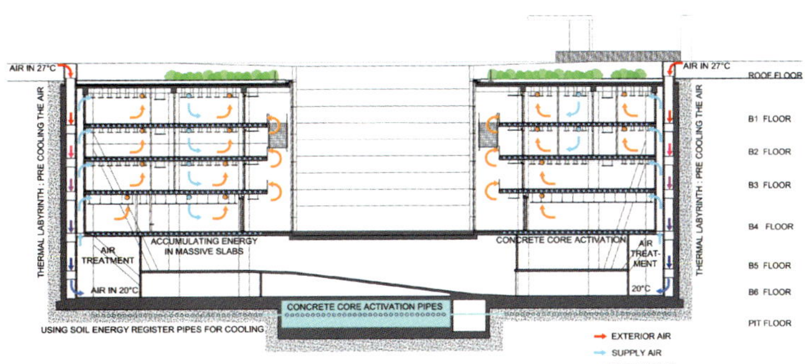

COOLING AIR CIRCULATION PLAN(SUMMER)

HEATING AIR CIRCULATION PLAN(WINTER)

The canopy of elevator box

A metalic object in sunken garden

Upper view seen sunken garden

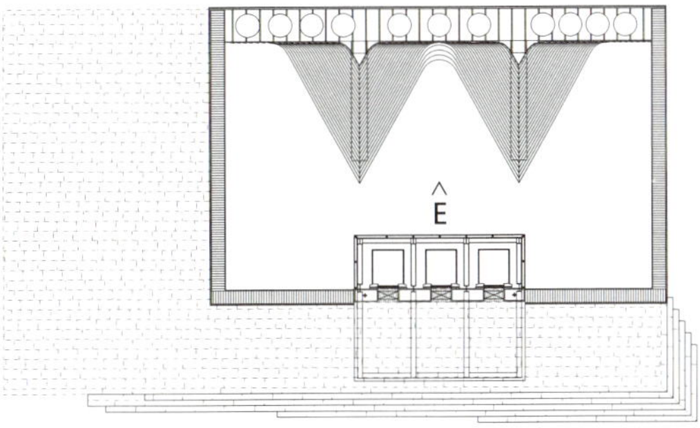

ELEVATOR ENTRANCE & SUNKEN GARDEN PLAN

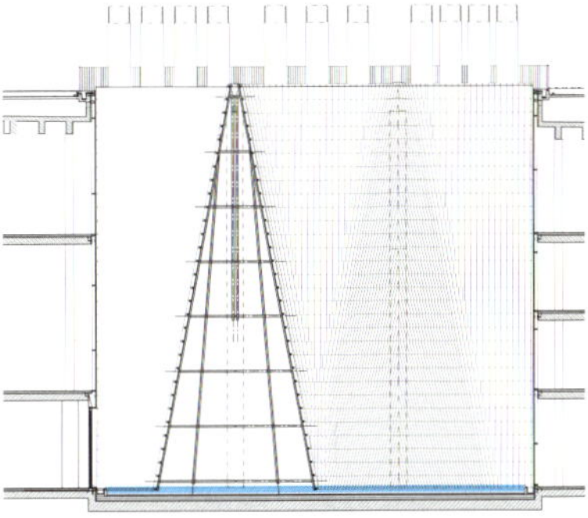

SUNKEN GARDEN METAL WALL ELEVATION E

>> COMPETITION_EWHA CAMPUS COMPLEX

Dominique Perrault Architecture Dominique Perrault

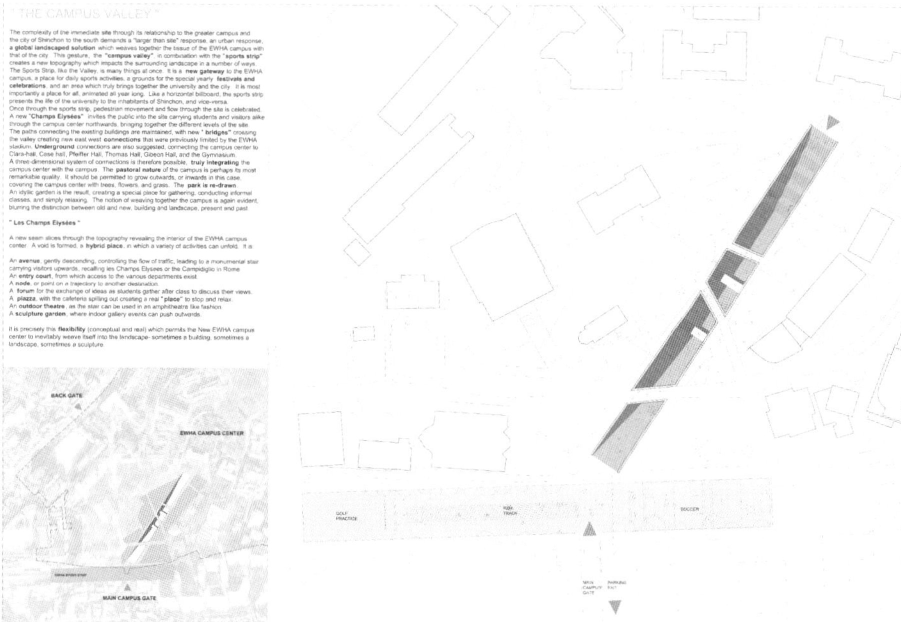

" THE CAMPUS VALLEY "

The complexity of the immediate **site** through its relationship to the greater campus and the city of Shinchon to the south demands a "larger than site" response, an urban response, **a global landscaped solution** which weaves together the tissue of the EWHA campus with that of the city. This gesture, the **"campus valley"**, in combination with the **"sports strip"** creates a new topography which impacts the surrounding landscape in a number of ways. The Sports Strip, like the Valley, is many things at once. It is a **new gateway** to the EWHA campus, a place for daily sports activities, a grounds for the special yearly **festivals and celebrations**, and an area which truly brings together the university and the city. It is most importantly a place for all, animated all year long. Like a horizontal billboard, the sports strip presents the life of the university to the inhabitants of Shinchon, and vice-versa.
Once through the sports strip, pedestrian movement and flow through the site is celebrated. A new **"Champs Elysées"** invites the public into the site carrying students and visitors alike through the campus center northwards, bringing together the different levels of the site. The paths connecting the existing buildings are maintained, with new **" bridges"** crossing the valley creating new east west **connections** that were previously limited by the EWHA stadium. **Underground** connections are also suggested, connecting the campus center to Clara-hall, Case hall, Pfeiffer Hall, Thomas Hall, Gibeon Hall, and the Gymnasium.
A three-dimensional system of connections is therefore possible, **truly integrating** the campus center with the campus. The **pastoral nature** of the campus is perhaps its most remarkable quality. It should be permitted to grow outwards, or inwards in this case, covering the campus center with trees, flowers, and grass. The **park is re-drawn**.
An idyllic garden is the result, creating a special place for gathering, conducting informal classes, and simply relaxing. The notion of weaving together the campus is again evident, blurring the distinction between old and new, building and landscape, present and past.

" Les Champs Elysées "

A new seam slices through the topography revealing the interior of the EWHA campus center. A void is formed, a **hybrid place**, in which a variety of activities can unfold. It is:

An **avenue**, gently descending, controlling the flow of traffic, leading to a monumental stair carrying visitors upwards, recalling les Champs Elysees or the Campidiglio in Rome.
An **entry court**, from which access to the various departments exist.
A **node**, or point on a trajectory to another destination.
A **forum** for the exchange of ideas as students gather after class to discuss their views.
A **piazza**, with the cafeteria spilling out creating a real **"place"** to stop and relax.
An **outdoor theatre**, as the stair can be used in an amphitheatre like fashion.
A **sculpture garden**, where indoor gallery events can push outwards.

It is precisely this **flexibility** (conceptual and real) which permits the New EWHA campus center to inevitably weave itself into the landscape- sometimes a building, sometimes a landscape, sometimes a sculpture.

DOMINIQUE PERRAULT ARCHITECTE EWHA WOMANS UNIVERSITY CAMPUS CENTER PROJECT "CAMPUS VALLEY" JANUARY 2004

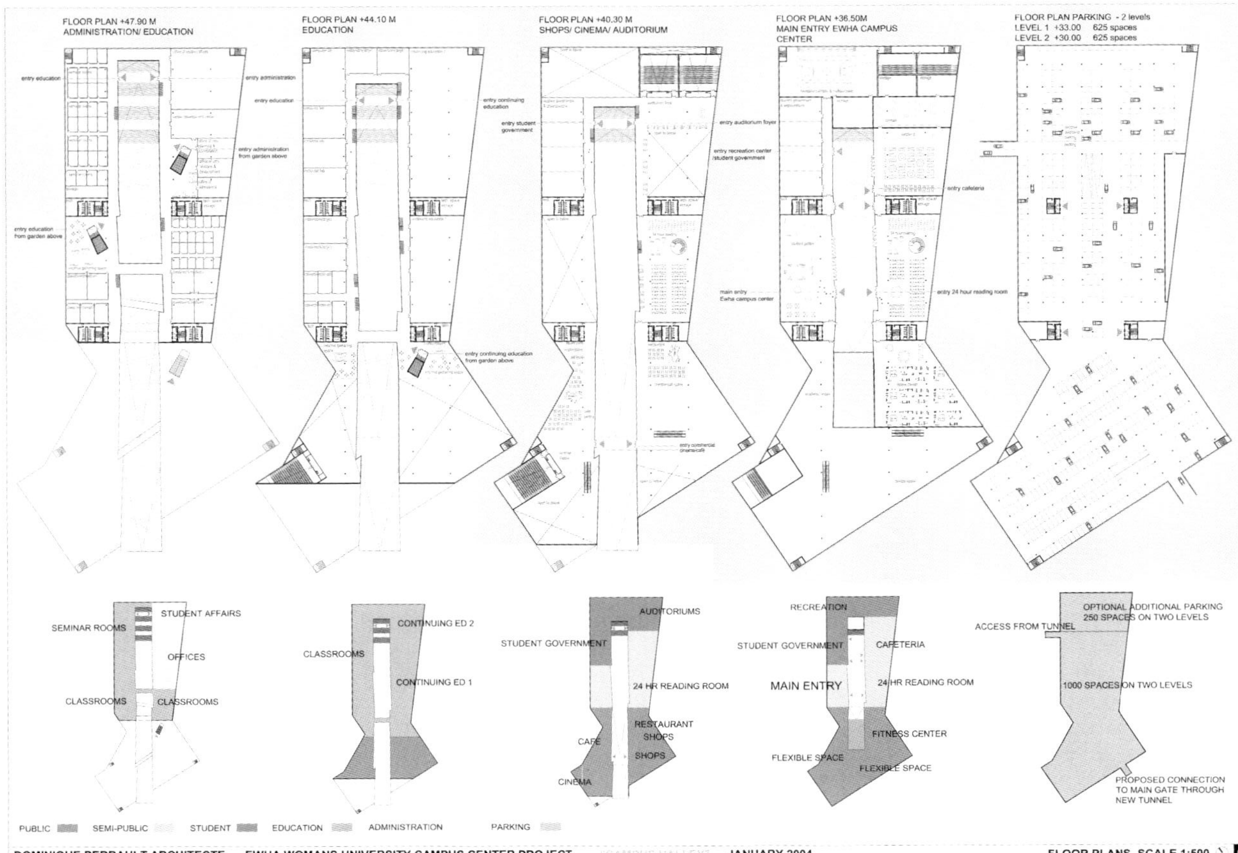
FLOOR PLAN +47.90 M
ADMINISTRATION/ EDUCATION
entry education
entry administration
entry education
entry administration from garden above
entry education from garden above
FLOOR PLAN +44.10 M
EDUCATION
entry continuing education
entry continuing education from garden above
FLOOR PLAN +40.30 M
SHOPS/ CINEMA/ AUDITORIUM
entry student government
entry auditorium foyer
entry recreation center /student government
main entry Ewha campus center
entry commercial cinema/café
FLOOR PLAN +36.50M
MAIN ENTRY EWHA CAMPUS CENTER
entry cafeteria
entry 24 hour reading room
FLOOR PLAN PARKING - 2 levels
LEVEL 1 +33.00 625 spaces
LEVEL 2 +30.00 625 spaces
SEMINAR ROOMS
STUDENT AFFAIRS
OFFICES
CLASSROOMS
CLASSROOMS
CLASSROOMS
CONTINUING ED 2
CONTINUING ED 1
AUDITORIUMS
STUDENT GOVERNMENT
24 HR READING ROOM
RESTAURANT
SHOPS
CAFE
SHOPS
CINEMA
RECREATION
STUDENT GOVERNMENT
CAFETERIA
MAIN ENTRY
24 HR READING ROOM
FITNESS CENTER
FLEXIBLE SPACE
FLEXIBLE SPACE
ACCESS FROM TUNNEL
OPTIONAL ADDITIONAL PARKING
250 SPACES ON TWO LEVELS
1000 SPACES ON TWO LEVELS
PROPOSED CONNECTION
TO MAIN GATE THROUGH
NEW TUNNEL
PUBLIC
SEMI-PUBLIC
STUDENT
EDUCATION
ADMINISTRATION
PARKING
DOMINIQUE PERRAULT ARCHITECTE
EWHA WOMANS UNIVERSITY CAMPUS CENTER PROJECT
"CAMPUS VALLEY"
JANUARY 2004
FLOOR PLANS SCALE 1:500
02

SECTION AA

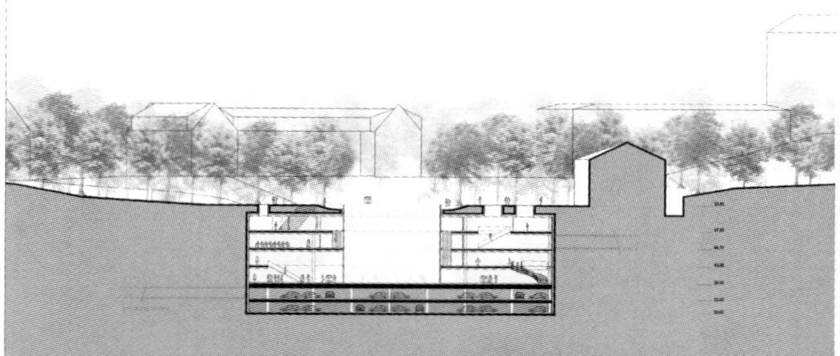
SECTION BB

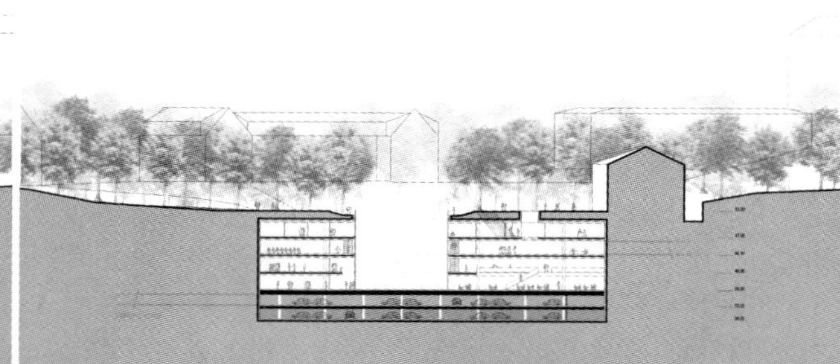
SECTION CC

DOMINIQUE PERRAULT ARCHITECTE **EWHA WOMANS UNIVERSITY CAMPUS CENTER PROJECT** "CAMPUS VALLEY" **JANUARY 2004** **SECTIONS** **03**

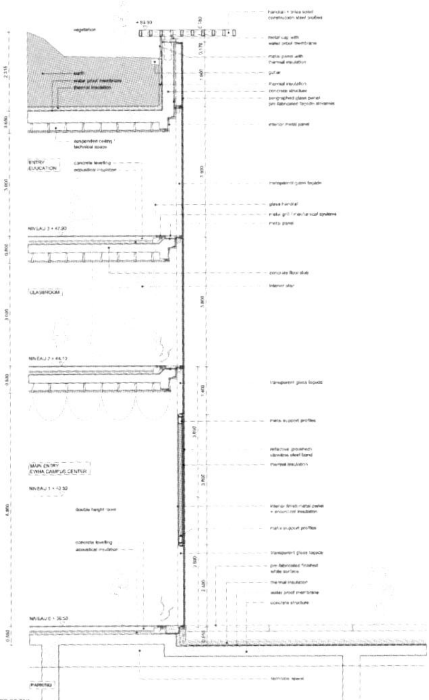

FACADE DETAIL

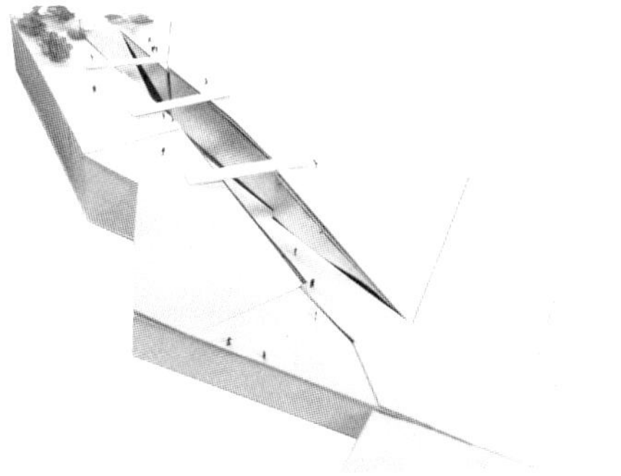

STUDY MODEL

Michael Heizer

Lucio Fontana

UIJEONGBU SCIENCE LIBRARY

Location Singok-dong, Uijeongbu-si, Gyeonggi-do, Korea **Site Attribute** Natural Green, Park **Site Area** 9,386.00m² **Building Area** 1,660.47m² **Total Floor Area** 6,951.70m²(4,986.76m²) **Landscape Area** 3,250.84m² **Building Coverage Ratio** 17.69% **Floor Area Ratio** 53.13% **Building Scope** B1~3F **Structure** RC **Exterior Finish** High-Density Wood Panel(WTC), T24 Clear Pattern Glass(KCC), T24 Color Pair Glass(KCC) **Interior Finish** Marble, Interior Film, MDF, Backpainted Glass **Design Period** 2004.11.5~2005.3.4 **Construction Period** 2005.9~2007.6 **Architecture Design** Yoon Sang-guk | KWAN Space Design Group Architects & Engineers **Design Team** Lee Sang-cheol, Lee U-seok, Kim Hyeon-deok, Yu Hwa-sun, An Jeong-su, Won Tae-hui, Jin Seung-hyeon **Construction** WOOIL Construction Industrial Co., Ltd. **Photographer** Lee Ki-hwan

Library viewed from the north-western entry road

PLAN FOR ARRANGEMENT With consideration of the shape of the site and the surroundings, the interconnected mass was planned to secure cognizability and accessibility. The spaces were arranged based on its function and interactions with others. While creating environment-friendly spaces internally, the architect focused on designing clear and functional exterior spaces with sufficient capacity.

PLAN FOR PLANES In the first under ground floor where visitors mainly enter to park their cars, there is a spacious lounge hall to accept a large number of visitors. Centering the hall, a culture classroom near the court and an performance space connected to the exterior on the east are placed. The inner court and the lounge hall, and a pond on the east are all linked through a bridge. Restaurants are arranged on the northern facade where can be kept out of the direct rays of the sun while data room and a library are placed on the western section where is easily accessible for lending books and the machinery room is placed on the southern underground floor. The first ground floor was designed to enable visitors to enter through the main entrance and move to the information center and the book cafe through the exhibition hall. The library for eldery people and disabled people was constructed within the northern section, the nearest place from the entrance. And family library and children's study were arranged within the green western section. Offices were arranged in the eastern section, and data room was built between the southern section and he central court where is linked with the lounge deck. When ascending to the second floor, we can see a reference room and publication center, as well as the special event corner at the center. In the informative multi-media room, one can enjoy upgraded learning experience. The general study room and a seminar room on the third floor are conveniently connected to the exhibition hall, adventure room, and observatory room as a scientific network.

PLAN FOR SECTIONS Environment-friendly materials were used to the building to make a great harmony with the surroundings. And oblique angles for the glass curtain walls were designed to represent future-going images. Organizational construction, functional partitioning, and energy-efficient volume and sections are the features of this building.

1 LIBRARY
2 PARKING LOT
3 ARTIFICIAL POND
4 OBSERVATION DECK
5 ECOLOGY POND
6 STAGE
7 SHELTER
8 GARDEN
9 MINERAL SPRING
10 BICYCLE DEPOSITORY

SITE PLAN

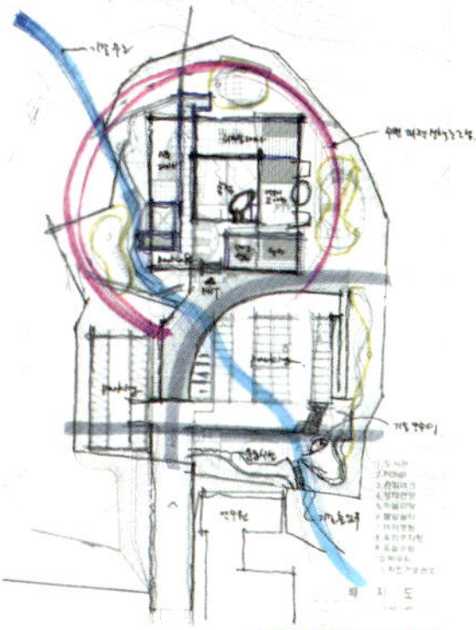

CONCEPT SKETCH

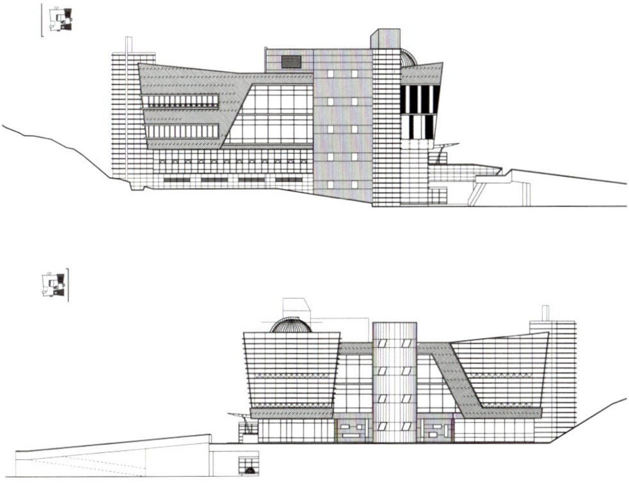

ELEVATION

Main entrance of library

Structure of entrance for car

Front part

South-western rear elevation

Exterior detail finished with wood panel

1 MACHINE ROOM
2 LIBRARY
3 PERFORMANCE HALL
4 PLAZA
5 CAFETERIA
6 CLASSROOM
7 DATA ROOM
8 ELECTRIC ROOM
9 GENERATOR ROOM
10 EMERGENCY ROOM
11 OFFICE
12 LEADING ROOM
13 DIRECTOR'S ROOM
14 BOOK CAFE
15 OUTDOOR RESTING ROOM
16 LANGUAGE LABORATORY
17 COMMUNICATIONS ROOM
18 MULTIMEDIA ROOM
19 RESTING ROOM
20 STUDY ROOM
21 SCIENCE LABORATORY

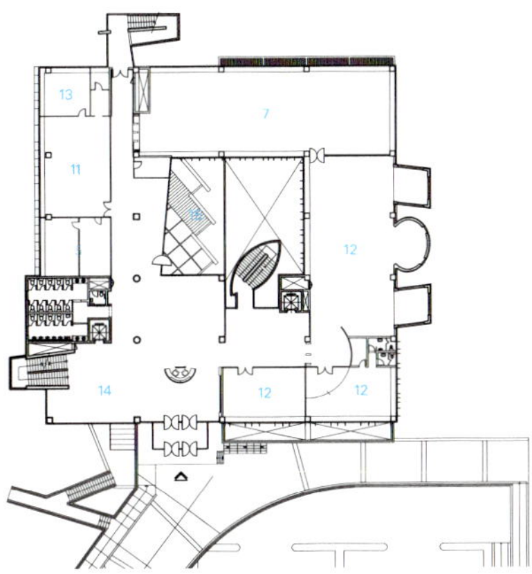

1ST FLOOR PLAN

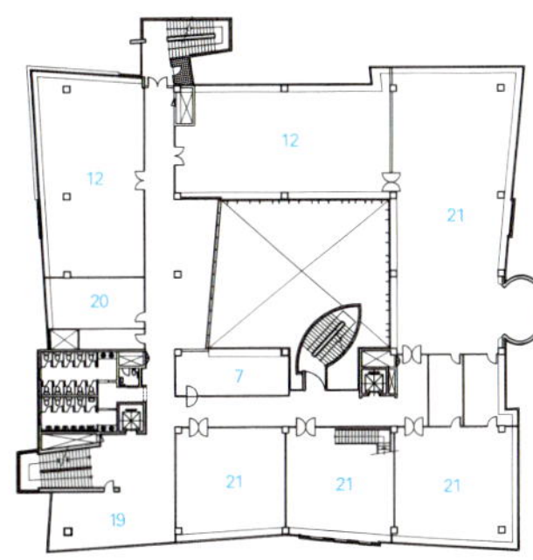

3RD FLOOR PLAN

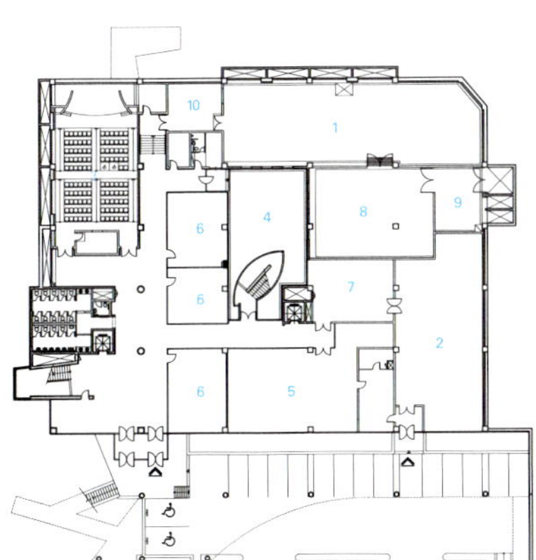

B1 FLOOR PLAN

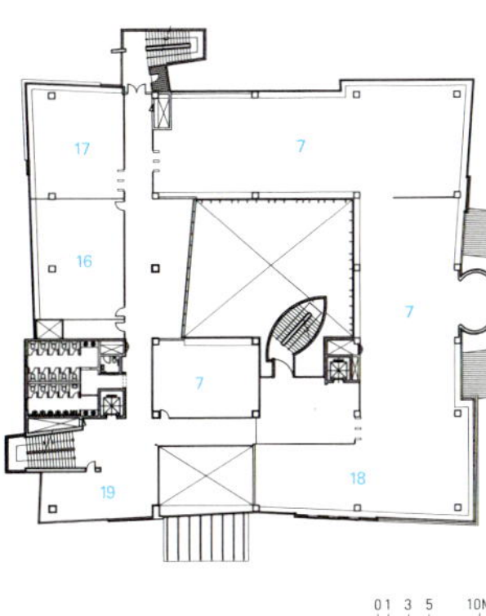

2ND FLOOR PLAN

Outdoor resting room on the 1st floor

Resting area on the 3rd floor

1 SCIENCE LAB
2 LITERATURE DATA ROOM
3 READING ROOM
4 STUDY OFFICE
5 KITCHEN
6 DATA READJUST ROOM
7 ELECTRIC ROOM
8 MACHINE ROOM
9 DIGITAL DATA ROOM
10 HVAC ROOM
11 OPERATING ROOM
12 COMPUTER ROOM
13 CONTROL ROOM
14 CLASSROOM
15 LIBRARY

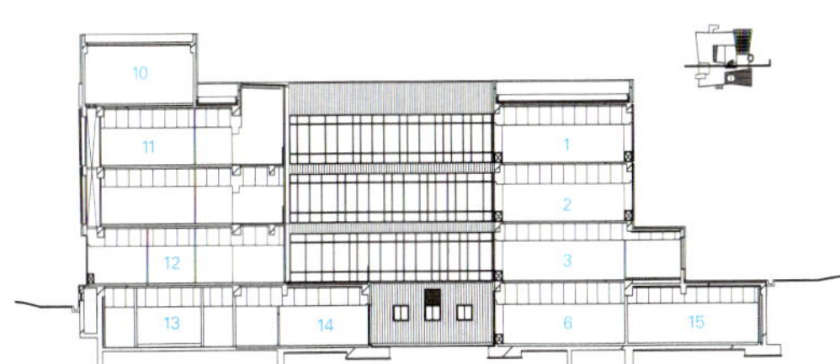

SECTION

Lobby on the 1st floor

Leading room on the 1st floor

Multimedia room on the 2nd floor

Exhibition hall on the 3rd floor

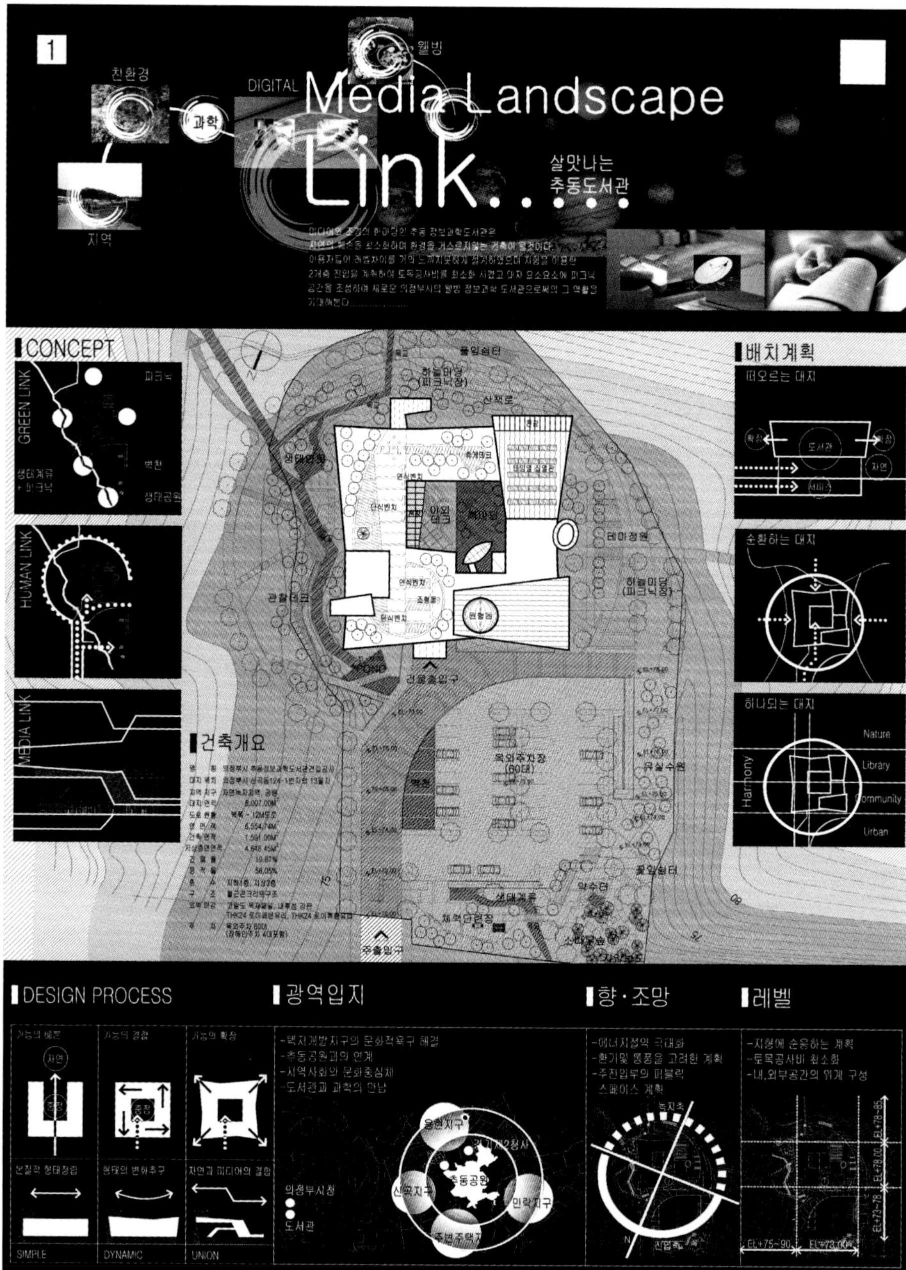
1
천환경
과학
웰빙
지역
DIGITAL Media Landscape
Link.....
살맛나는
추동도서관
CONCEPT
GREEN LINK
피크닉
벽천
생태계류
+ 피크닉
생태공원
HUMAN LINK
MEDIA LINK
배치계획
떠오르는 대지
도서관
자연
순환하는 대지
하나되는 대지
Nature
Library
Community
Urban
Harmony
하늘마당
(피크닉장)
산책로
테마정원
관찰데크
POND
건물출입구
건축개요
옥외주차장
(60대)
유실수원
꽃잎쉼터
약수터
체력단련장
주출입구
DESIGN PROCESS
SIMPLE
DYNAMIC
UNION
광역입지
-택지개발지구의 문화적욕구 해결
-추동공원과의 연계
-지역사회의 문화중심체
-도서관과 과학의 만남
의정부시청
도서관
용현지구
신곡지구
추동공원
민락지구
주변주택지
향·조망
-에너지절약 극대화
-환기및 통풍을 고려한 계획
-주진입부의 퍼블릭
스페이스 계획
녹지축
진입축
N
레벨
-지형에 순응하는 계획
-토목공사비 최소화
-내,외부공간의 위계 구성

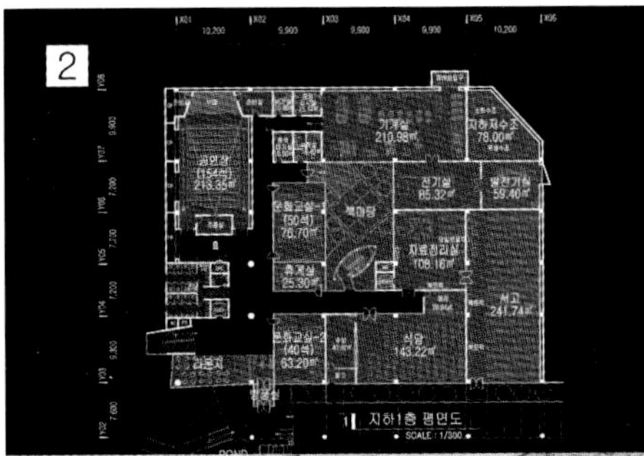

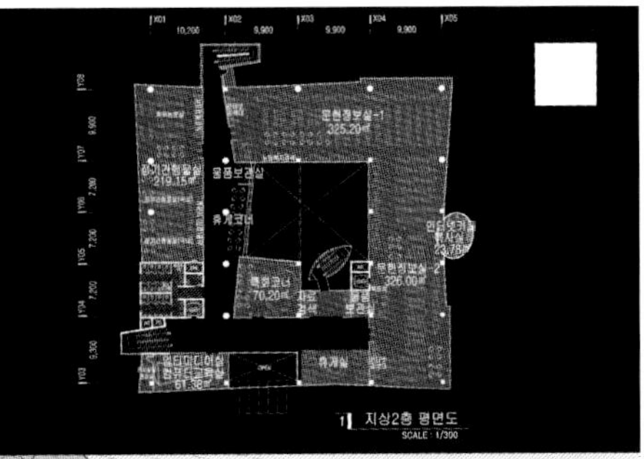

친환경계획

■ 태양열 시스템
- 대체 에너지의 저변 보급 및
  실용화 될수있는 에너지자원 확보

■ 수공간
- 간접증발로 인한 냉각효과 기대
  동식물의 비오톱 효과 기대.

■ 차양
- 수평비를 이용한 직사광 차단
  루버를 이용한 직사광 차단(아뜨리움)

도서관리 시스템 계획

■ 도서관 서비스 향상
- 자료의 대출및 반납에 소요되는
  대기시간의 최소화
- RFID시스템으로 인한 도서관
  서비스의 질적 향상

■ 도서관 업무효율 극대화
- 별도의 분실방지 태그 부착없이
  RFID 태그를 통한 분실방지
- 장서 점검이 빠르고 정확하게
  수행되어 업무능력 향상

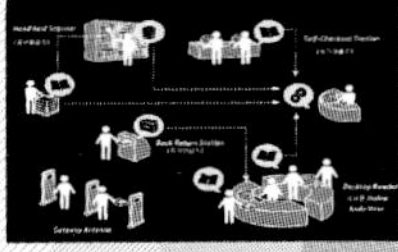

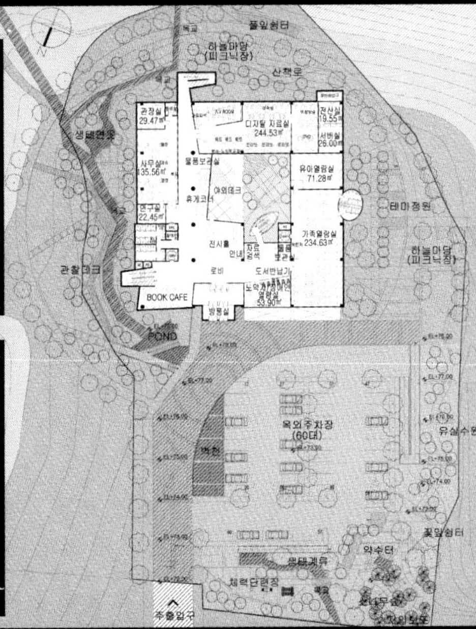

평면구성

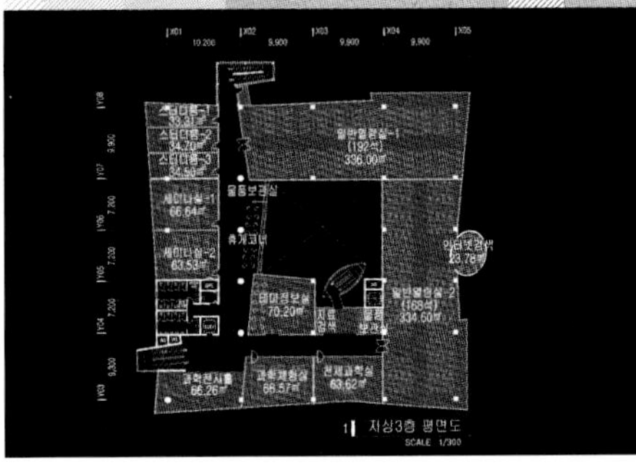

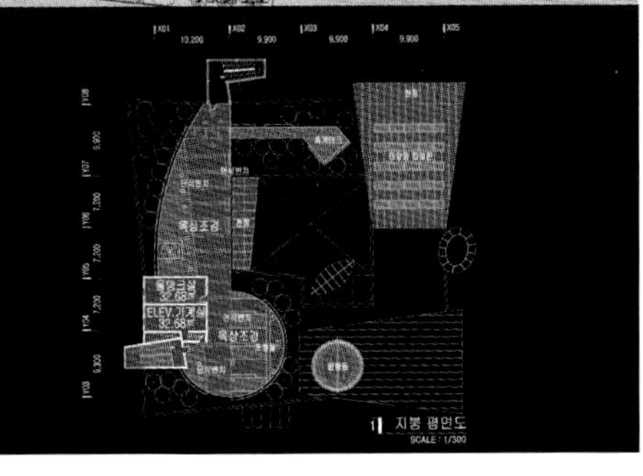

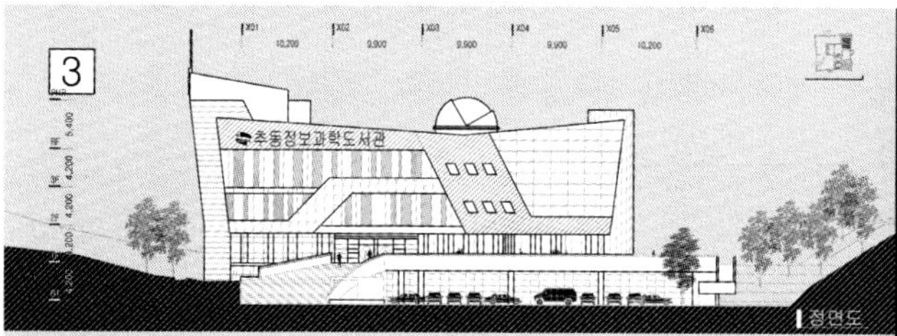

정면도

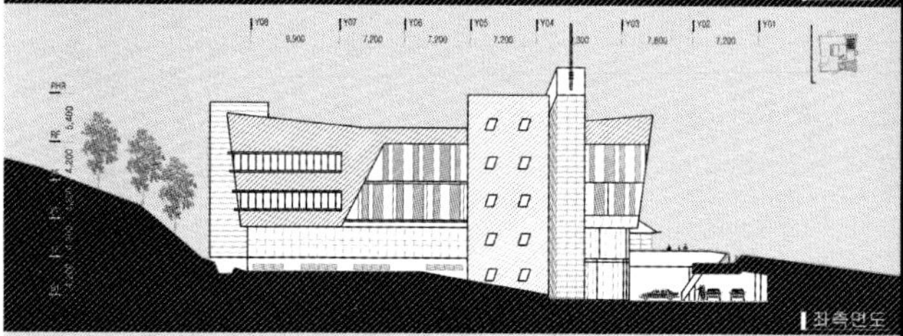

좌측면도

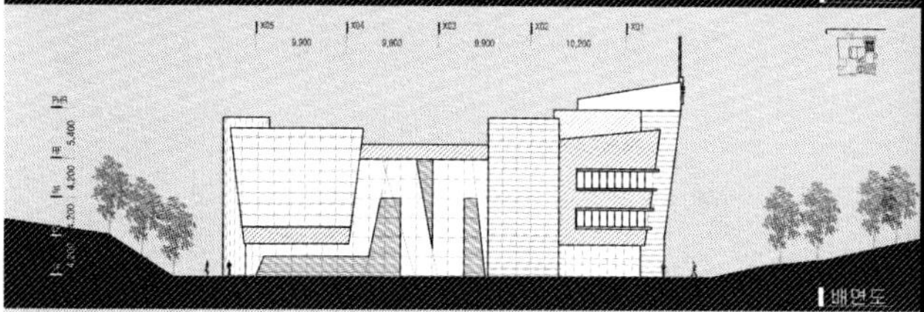

배면도

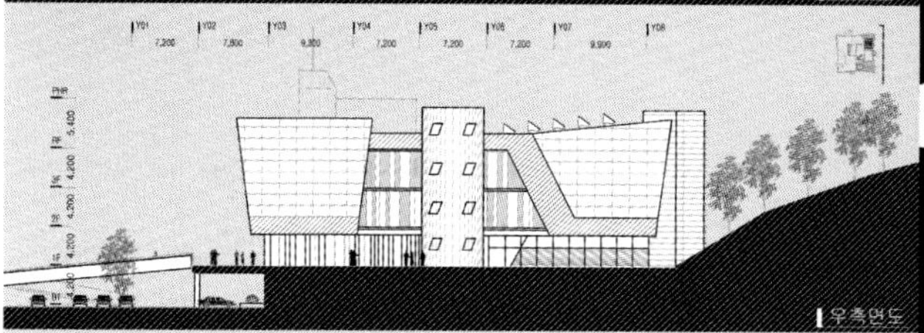

우측면도

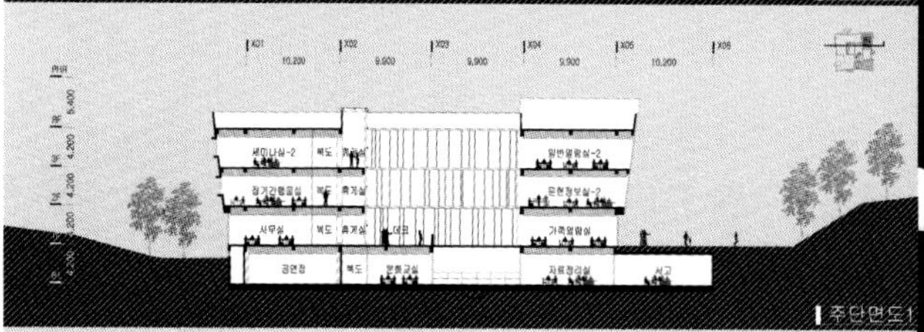

주단면도1

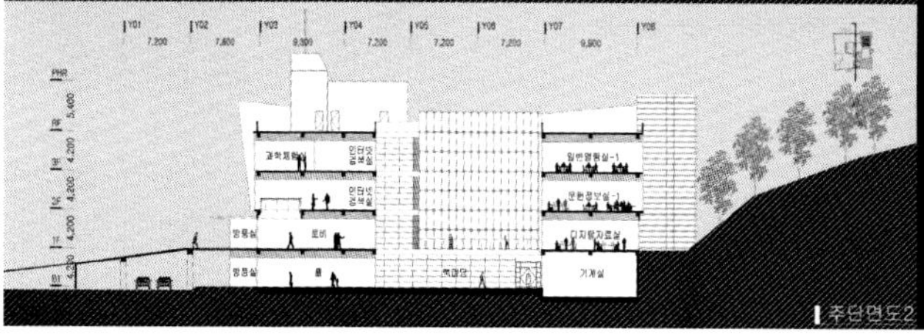

주단면도2

입면계획

떠오르기 (LANDMARK)

- 공공시설로써의 랜드마크적 이미지부여 빗각으로 올라가는 건물의 부양

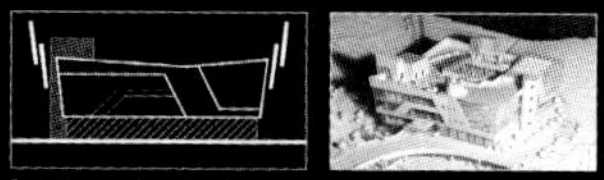

이어주기 (UNION)

- 자연의 관입과 삽입을 통한 주변환경과 건물매스와의 결합

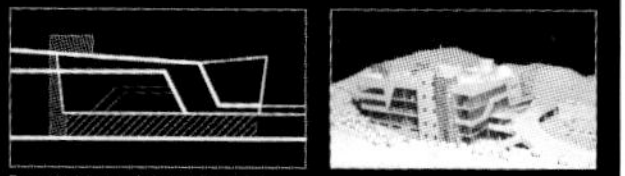

도약하기 (DYNAMIC)

- 주변 자연수림대와 어울리는 진취적인 기상 및 의정부시의 힘찬도약

단면계획

지형의 레벨차를 이용한 2개층 진입 및 필로티주차

- 지형차를 이용한 지하, 지상 출입구 확보
- 지형을 순응하고 절토, 성토를 최소화하는 단면계획

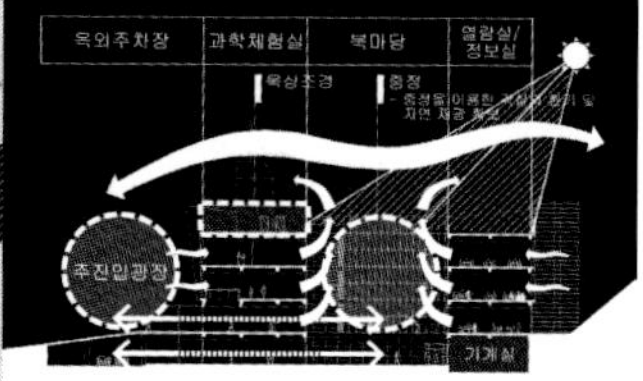

SUNKEN, 중정을 이용한 자연 환기.채광

- 지하층 공간의 환기, 채광을 위한 썬큰가든, 중정
- 에너지절약을 위한 각실의 채광 및 자연환기 확보(남,동향배치)

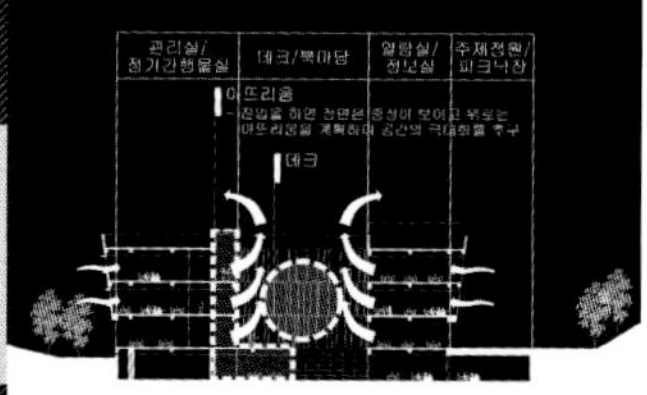

수직루버를 이용한 일광조절

- 수직 루버를 이용한 다양한 공간 분위기 연출
- 루버를 이용한 입체효과 연출

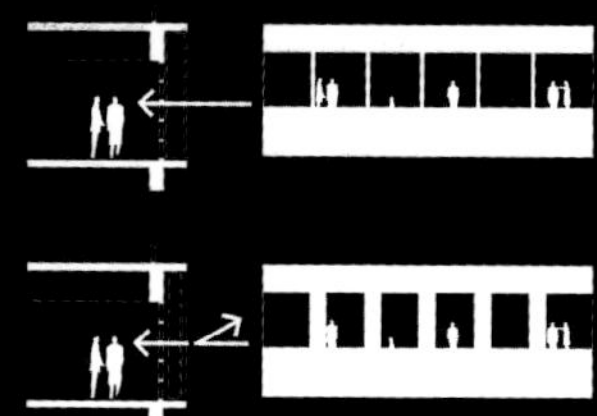

4

피크닉 조성계획

■ 기 본 방 향

- 자연 친화적인 공간조성
- 다양하고 즐거운 체험공간의 연속적 공간조성
- 연령층을 초월한 지역주민을 위한 옥외환경조성
- 문화 중심의 샘터로서의 옥외공간구성
- 다양한 이벤트(전시,전람,공연 등)가 함께 공존하는 문화공간 조성

■ 공 간 계 획

- 테마정원 (이벤트장, 전시,전람 문화공간, 야외소극장)
- 피크닉장 (잔디마당, 산책로, 쉼터)
- 생태연못 (생태계류, 연못, 관찰데크)
- 체력단련장 (체육시설, 지압보도)
- 약수터 (쉼터)
- 유실수원 (자연학습장)

조 경 계 획

■ 기 본 방 향

- 자연 친환경적 외부공간 조성
- 교육의 최적 환경조성
- 체험학습, 현장학습의 기회 제공의 교육환경 조성
- 쾌적하고 다양함이 어우러지는 문화와 쉼터공존의 외부공간 조성
- 계절의 변화를 느낄 수 있는 아름다운 공간 조성

■ 공 간 계 획

- 자연녹지공간의 보존과 녹지공간의 확대
- 각 공간의 장소성 부여

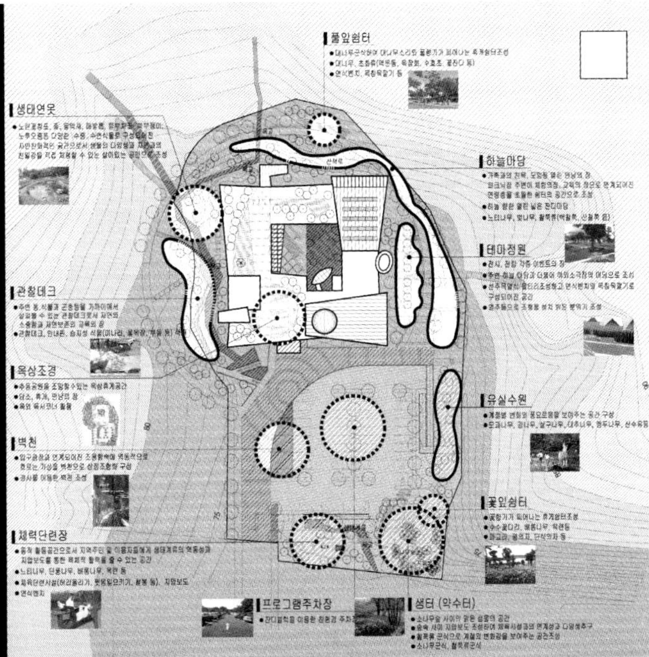

모형 사진

동선계획

ANSAN COLLEGE SEONGSIL BLDG

Location Il-dong, Sangnok-gu, Ansan-si, Gyeonggi-do, Korea **Site Attribute** Natural Green, School Facilities Protection **Site Area** 144,637.60m² **Building Area** 2,853.27m² **Total Floor Area** 8,726.14m² **Landscape Area** 69,851.14m² **Building Scope** B1~4F **Structure** RC, Part Steel **Exterior Finish** Asbestos Free Extruded Formed Cement Board, Ceramic Panel, Red Cedar, Color Pair Glass, Drybit Topcoat **Interior Finish** Granite, Marble Panel, Asbestos Free Extruded Formed Cement Board, Water-Based Paint **Design Period** 2005.10.2~2006.2.25 **Construction Period** 2006.2.25~2007.1.19 **Architecture Design** Kim Ho-joon, Park Young-goo | ADOS Design Group + Choi Gyu-hak | ECOLOGICAL Architecture & Urbandesign **Design Team** Bae Yeon-ju, Lee Sang-yeop, Sin Dong-jae, Sin Do-hyeon, Kim Do-hui, Lee Dong-ha, Yang Ji-eun **Construction** DAEYANG Construction **Client** Ansan College **Photographer** Lee Ki-hwan

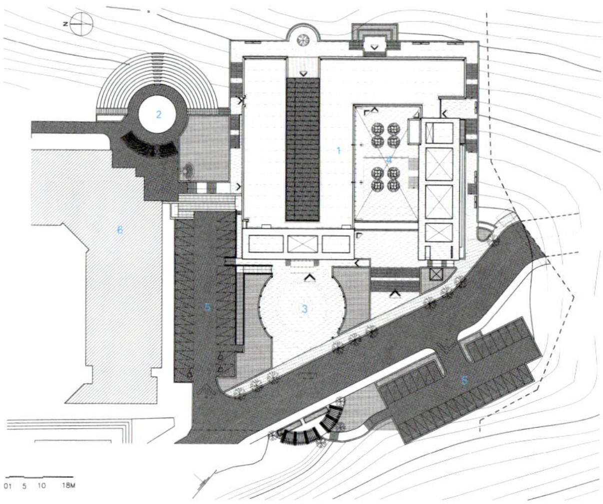

SITE PLAN

STUDY MODEL

1 SEONGSIL BUILDING
2 OUTDOOR STAGE
3 ENTRY SQUARE
4 SQUARE OF COMMUNICATION
5 PARKING LOT
6 EXISTING BUILDING

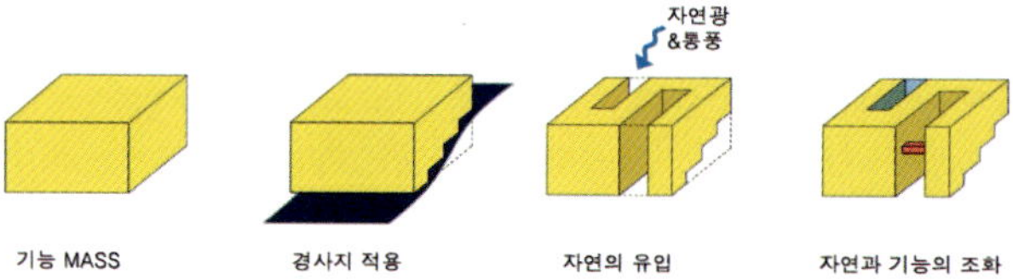

MASS PROCESS

The First Ansan College, which is located at a point where the city and the countryside meet, has an excellent environment in which the campus - except the main gate - is surrounded by low hills. In particular, Seongsil Hall was built on a site of 70mx70m positioned at the inner part of the campus. This project was designed with a few basic concepts under the condition that the difference of elevation should be more than 17m and the special design guideline must be met.

First, the building should be constructed in a way that it can reflect and realize the future picture of the college. Being jointly used by the Department of Nursing Science and Physical Therapy, the building should have enough flexibility to cope with the future development and changes in the school, and provide such an environment that students can receive high-quality educational services and professors work without inconveniences. A proper arrangement and combination of functions are required with the module and the direction fit for the function of each room. Thus, the building will have a future-oriented image as a whole through elevation responding to its surrounding environment.

Second, the building should be in good harmony with the natural environment. To do this, they created step-type paths in consideration of the inclined topography and installed devices communicating with its surrounding nature through a proper combination of traffic lines and outside spaces on each level. We considered the control of solar radiation and energy saving through change of elevation corresponding to each direction. The atrium will allow daylight into lecture rooms and laboratories and harmonize the living environment between inside and outside spaces.

Third, the building should be a space for communication between students and professors and among students themselves. Lecture rooms, laboratories and research rooms were separately placed but spaces for communication were provided on each level. The role of communication is expected in various outside spaces called Front Square, Square of Communication, Square of Thought and Garden of Meditation, with each of them on a different level.

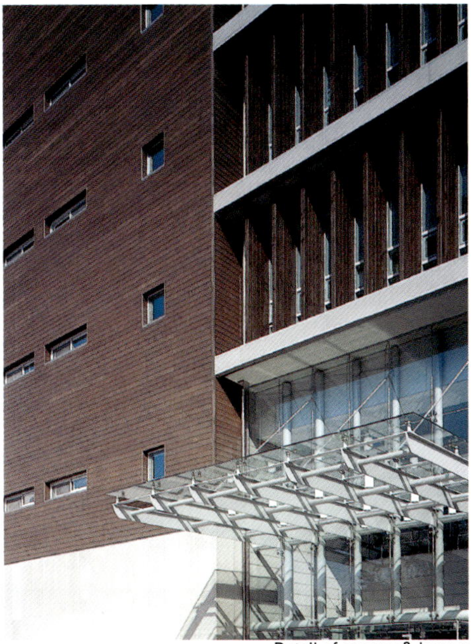

Detail of canopy & louver

Main entrance of lecture building(seongsil building)

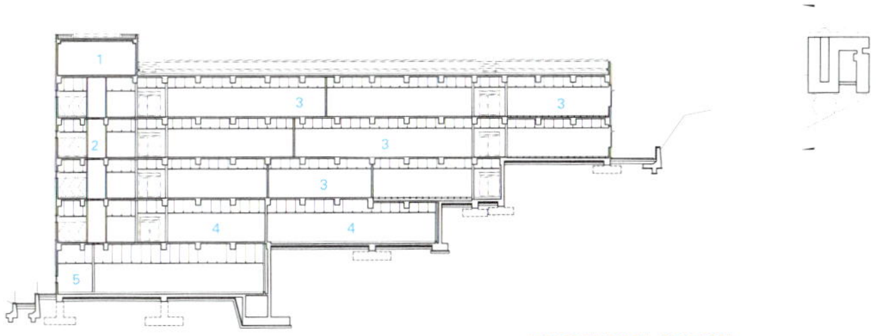

1 FAN ROOM
2 TOILET
3 PRACTICAL ROOM
4 LECTURE ROOM
5 RESTING ROOM
6 ATRIUM
7 LOBBY
8 PROFESSOR'S OFFICE
9 GENERATOR'S ROOM

LONGITUDINAL SECTION

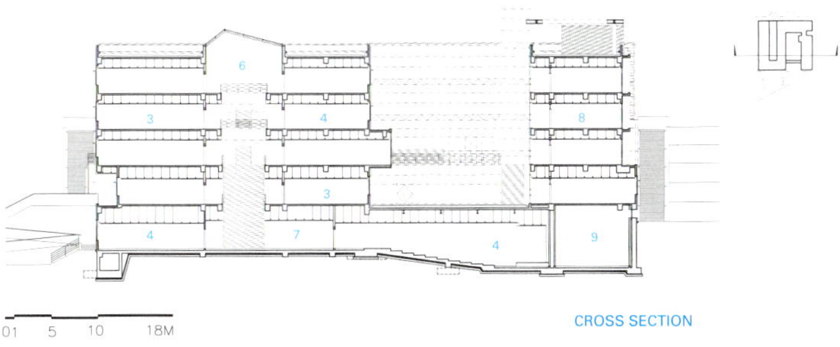

CROSS SECTION

The spaces encircling the plaza are linked to one another by the bridge.

Square of communication

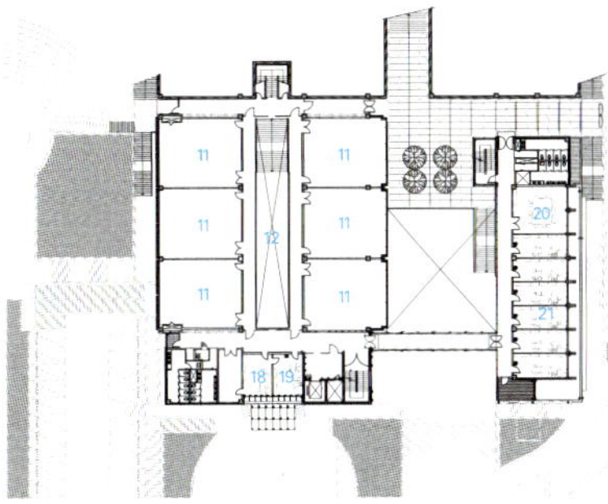

2ND FLOOR PLAN

1 LECTURE ROOM
2 LOBBY
3 CENTRAL STAIRS
4 STAFF'S RESTING ROOM
5 CONTROL ROOM
6 PIT
7 WATER TANK ROOM
8 MACHINE ROOM
9 GENERATOR ROOM
10 ELECTRIC ROOM
11 PRACTICAL ROOM
12 ATRIUM
13 HALL
14 SQUARE OF COMMUNICATION
15 LOUNGE
16 SNACK BAR
17 STORAGE
18 OFFICE
19 DIRECTOR'S ROOM
20 SEMINAL ROOM
21 PROFESSOR'S OFFICE
22 CIRCLE ROOM

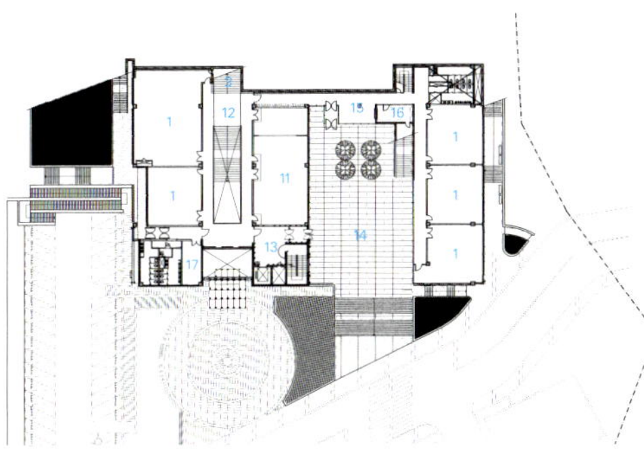

1ST FLOOR PLAN

4TH FLOOR PLAN

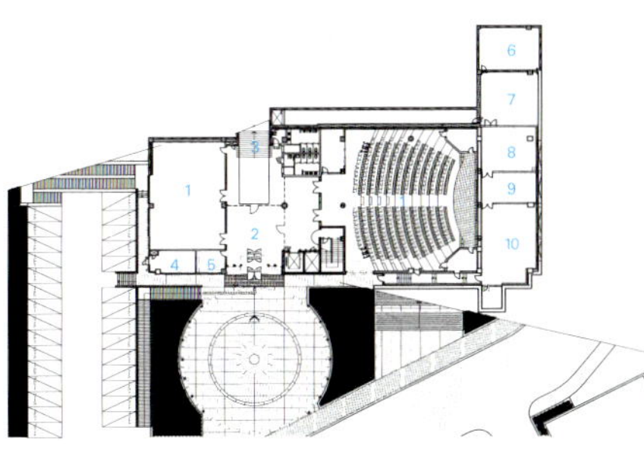

B1 FLOOR PLAN

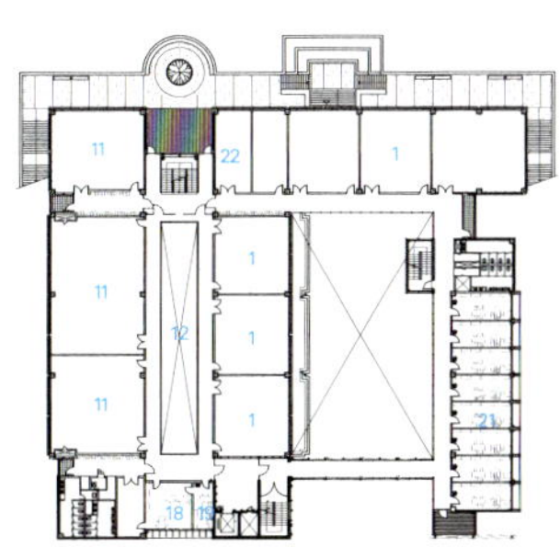

3RD FLOOR PLAN

Atrium open throughout five floors

Natural daylighting is introduced into the atrium through skylight.

1 안산1대학 강의동(성실관) 설계 계획안

P 250

■설계개요

명 칭	안산1대학 강의동(성실관)시설공사 건축설계경기
대지위치	경기도 안산시 상록구 일동 752, 752-1
지역지구	자연녹지지역, 보존녹지지역/학교시설 보호구역
도로면적	144,637.60㎡
도로현황	20m, 10m (계획도로)에 접함
연 면 적	57,857.19㎡ (신청면적 : 8,237.82㎡)
건축면적	15,930.42㎡ (신청면적 : 2,794.48㎡)
건 폐 율	11.01% (15,930.42/144,637.60×100)
용 적 률	34.91% (50,496.25/144,637.60×100)
구 조	철근 콘크리트조 / 일부 철골조
층 수	지하1층, 지상4층
최고높이	24.75m
외부마감	세라믹타일(건식공법), 적삼목, 노출콘크리트(실리콘뿜칠)
설비개요	냉난방 : EHP방식, 급탕 : 태양열 급탕시스템
주차개요	설계 : 311대 (법정 : 287대)

■ 모형사진

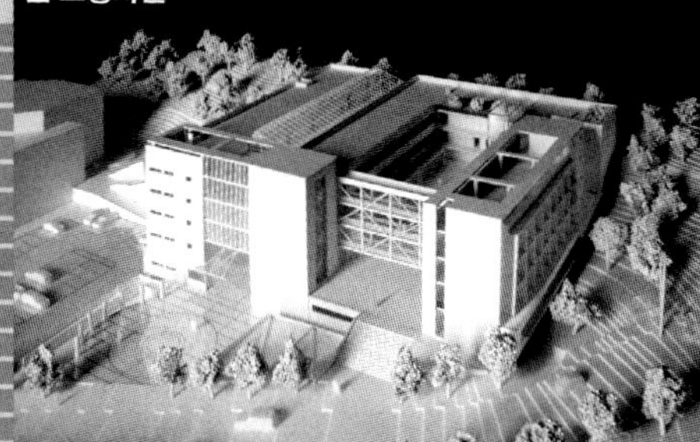

■배 치 도 SCALE 1/300

■대지 , 지형

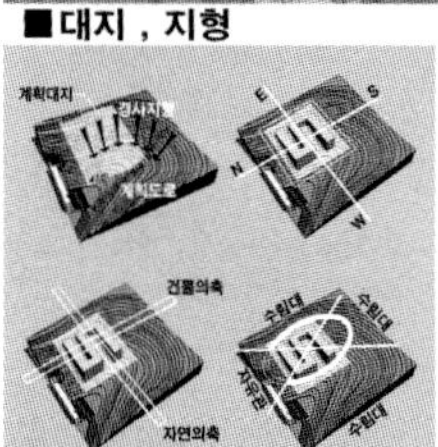

■형태개념

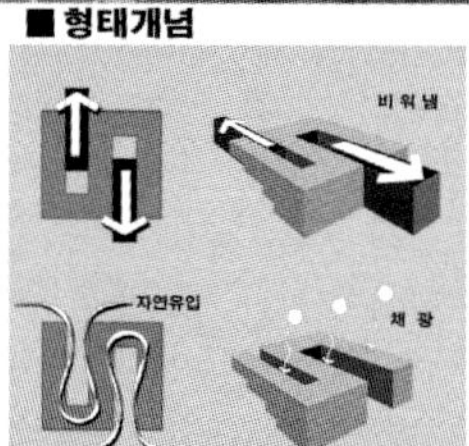

■대지이용

■동선및조경계획

052

EDUCATION
ANSAN COLLEGE SEONGSIL BLDG

3

P 250

■기능배치

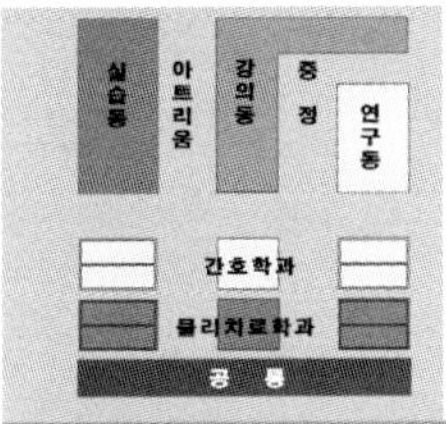

■모듈계획

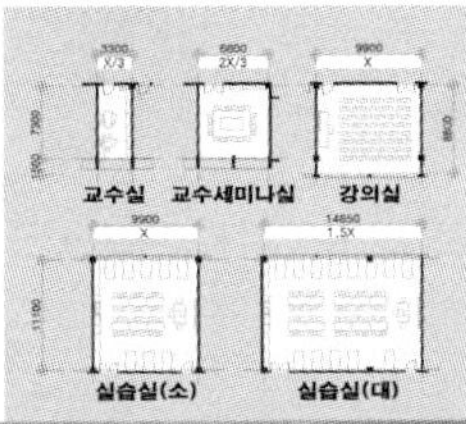

■단면계획

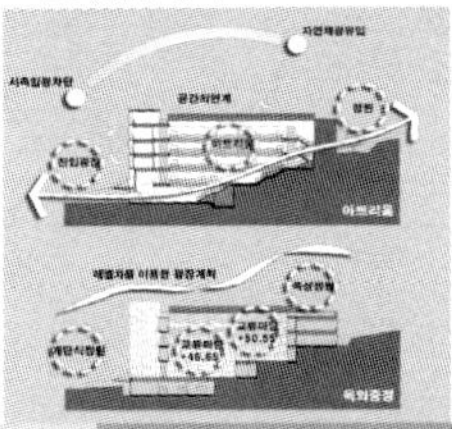

■형태구성

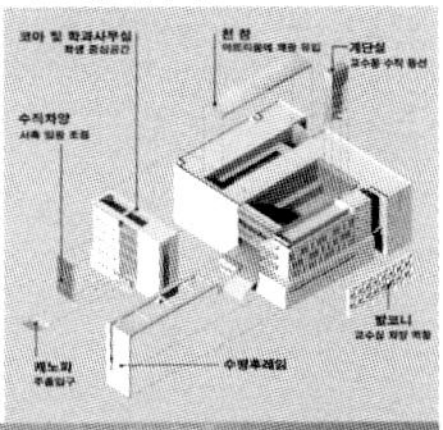

■입면도 SCALE 1/300

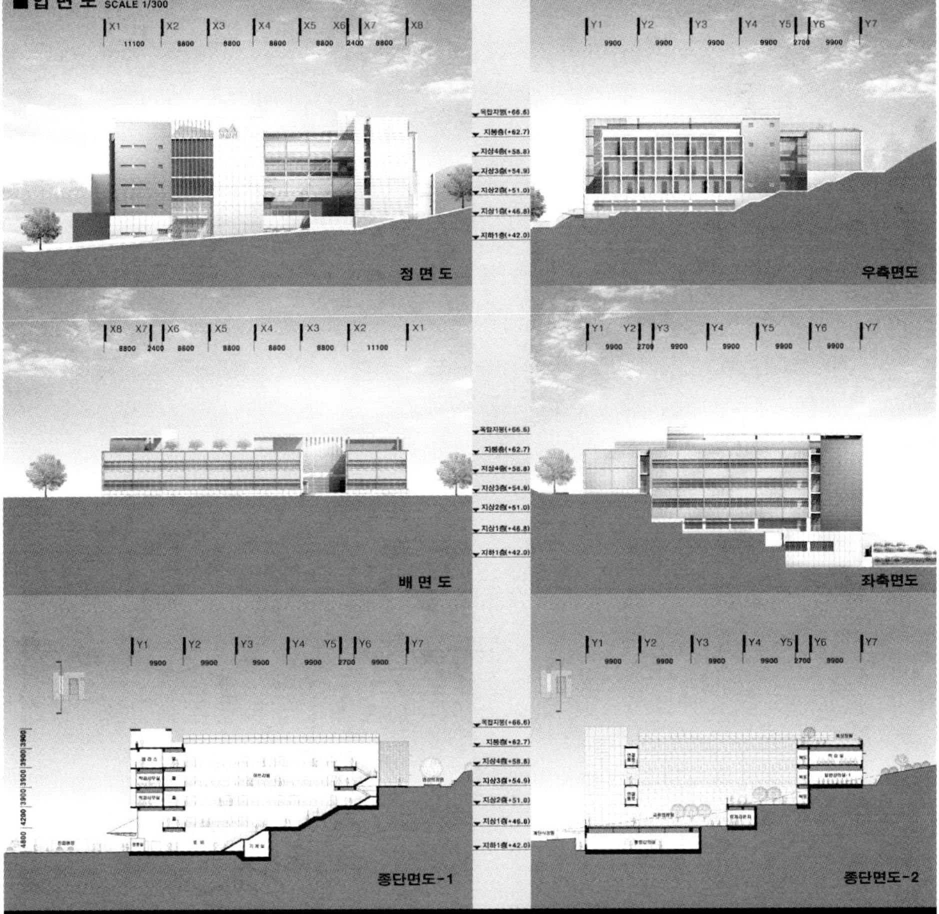

SAMSUNG LIBRARY OF SKKU

Location Cheoncheon-dong, Jangan-gu, Suwon-si, Gyeonggi-do, Korea **Site Attribute** City, General Resident, School Facilities Protection **Site Area** 492,312.5m² **Building Area** 4,159.75m² **Total Floor Area** 23,742.01m² **Building Coverage Ratio** 0.84% **Floor Area Ratio** 3.34% **Building Scope** B2~7F **Maximum Height** 38.1m **Structure** SRC **Parking** 131 Cars **Construction Period** 2009.3 **Architecture Design** Sohn Myung-gi, Koh Kwang-seog | SAMOO Architects & Engineers **Design Team** Whang Kyung-mee, Nam Ik-ho, You In-hee **Photographer** Lee Ki-hwan

West elevation

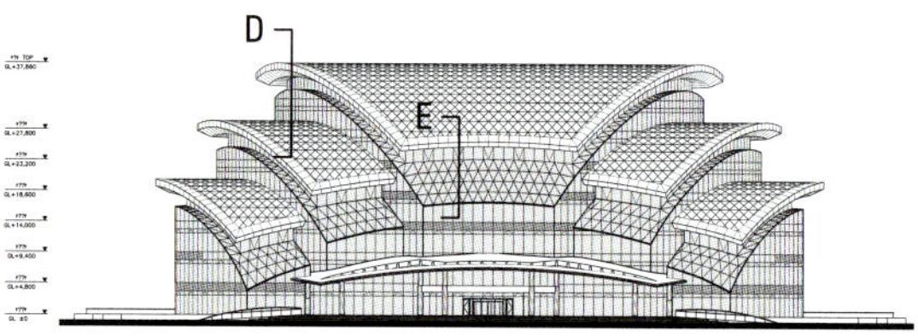

FRONT ELEVATION

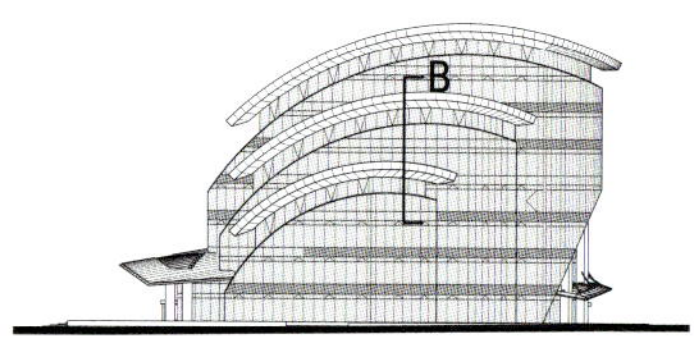

RIGHT ELEVATION

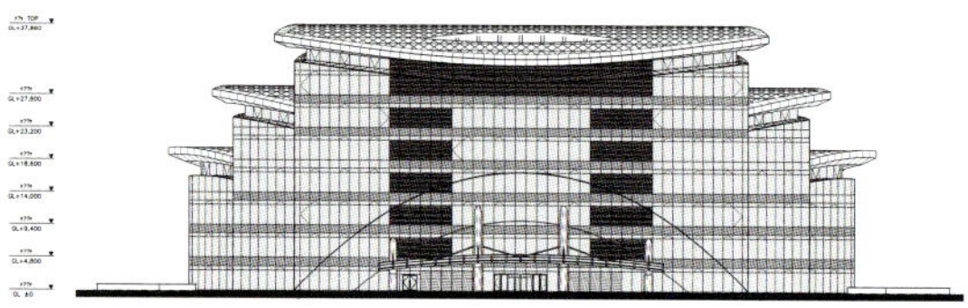
REAR ELEVATION

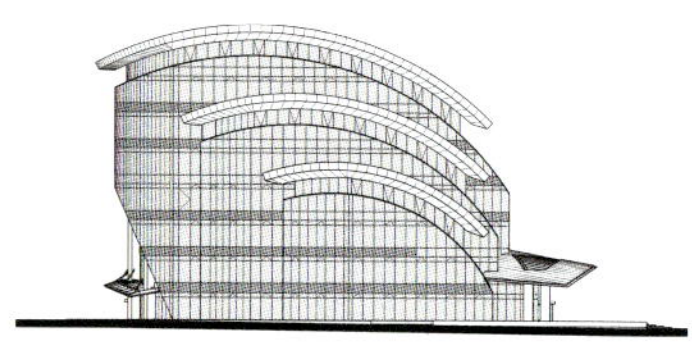
LEFT ELEVATION

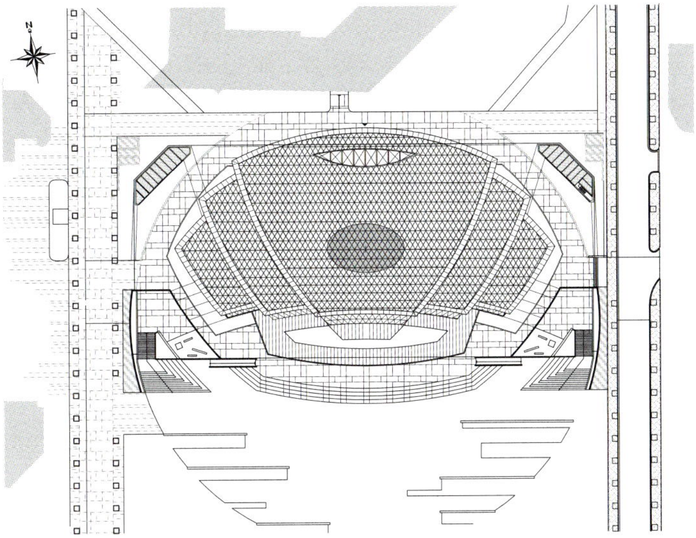

SITE PLAN

Waves of rapidly changing information in the digital era and overflowing information makes it necessary to systemize information and prepare spaces to provide and receive information to take the lead in this era and cope with various needs of users. The key points in the design are the establishment of identity of a scientific information center that takes a role of the core building in the campus, places to exchange information of the digital era, study spaces with new concepts, spaces for students to stay and utilize for various cultural activities, a library, an internet cafe, a multi-purpose complex space with a rest room and convention function and the expression of digital images from future-oriented high-tech information facilities.

The theme of a 'digital library' became the most basic concept of the design and continuous images of an open book express an open space with flowing information. Also the growth curve of a gingko leave, which is the symbol of Sungkyunkwan University was applied on the front side to suggest the motive from a gingko leave. The digital library that was expressed with transparent metallic materials in accordance with modern high-tech facilities will re-organize the campus context as a hub for information exchange in the center of the campus. Inside of the digital scientific information center, bright sunlight is induced through a big hall and a conference room that is more projected than other places expresses the vitality of dynamic and energetic students. Conference rooms in various shapes and sizes connect spaces that could easily separated from other floors and offer dynamic fun from architectural spaces to students. Large spaces that are opened with bilateral symmetry secure a wide range of views and naturally lead users' traffic lines.

The reading space in the lower part is classified into a free reading zone. Unlike normal libraries, a free reading zone is operated for 24 hours in an autonomic way. Lighting inside of the building was maximized by using sunken to create pleasant atmosphere and internal and external traffic lines were separated to stress natural introduction of traffic lines from outside. Multimedia zones on the 1st and 2nd floors are expected to record the biggest usage by students. These multi-functional multimedia spaces include rooms for information search, audio and video materials, a data room, a conference room, a studio, a small size theatre with about 40 seats and a rest room. The 3rd and 4th floors where the existing central library was transferred in its original style are respectively divided into an open reading room and a free reading room. On the 5th floor is a free community zone that is a complex space in a new concept. It will not be a closed study room with no sound, no color andno allowed activity but more like a library in a new concept where students can search information with lap tops, listen to music with MP3, lean against a comfortable sofa holding a cup of coffee and discuss or study with their friends. The 6th and 7th floors consist of a space for staff, an information and communication room and a server room and there is an observatory to view the central hall in the open hall on the 7th floor. The digital scientific information center that is newly organized with various spaces in various concepts will be the central place in the campus master plan as well as the most widely used space by students owing to its various functions.

Text by Whang Kyung-mee(Samoo Architects & Engineers)

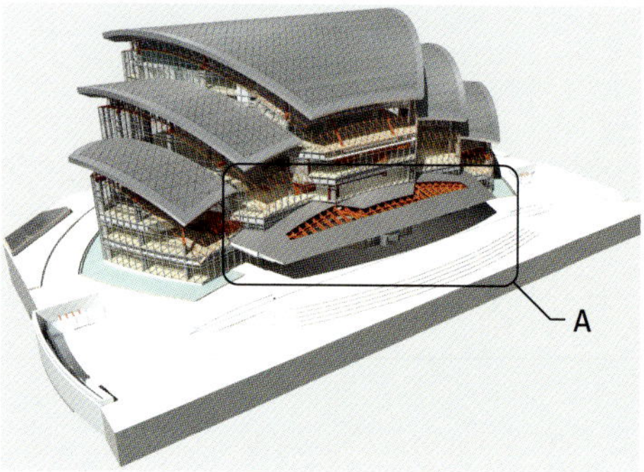

3D MODELING

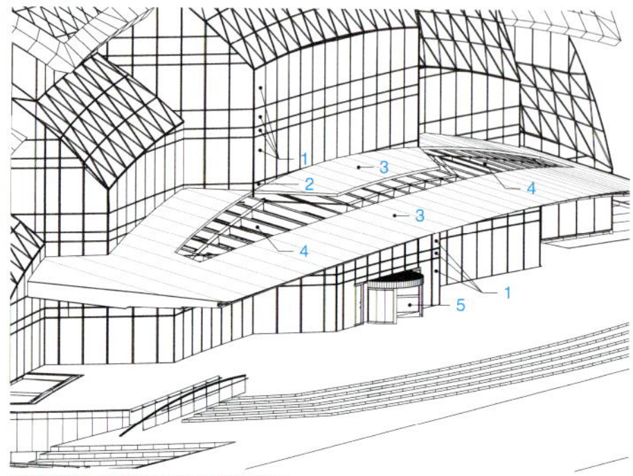

FRONT CANOPY AXONOMETRIC A

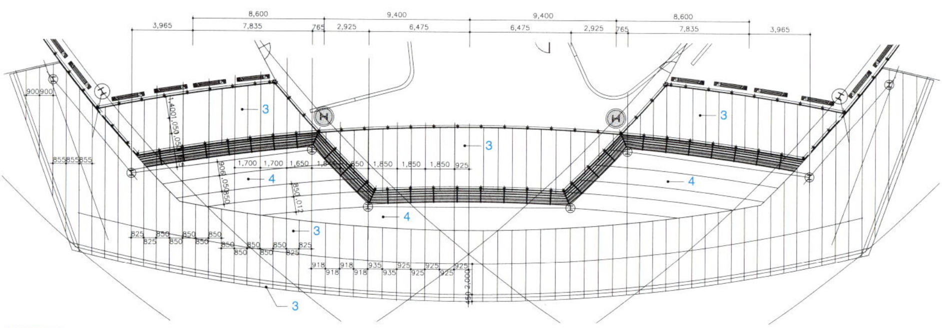

CANOPY PLAN

1 T24 PAIRED, ONE SIDE SEMI-TEMPERED GLASS
2 T24 PAIRED, TEMPERED GLASS(BOTH SIDE)
3 T3 AL. SHEET
4 T19.5 LAMINATED GLASS
5 T12 TEMPERED GLASS
6 ROOF 1
7 ROOF 2
8 ROOF 3
9 ROOF 4
10 ROOF 5
11 SRC 1A
12 SRC 2
13 SRC 3

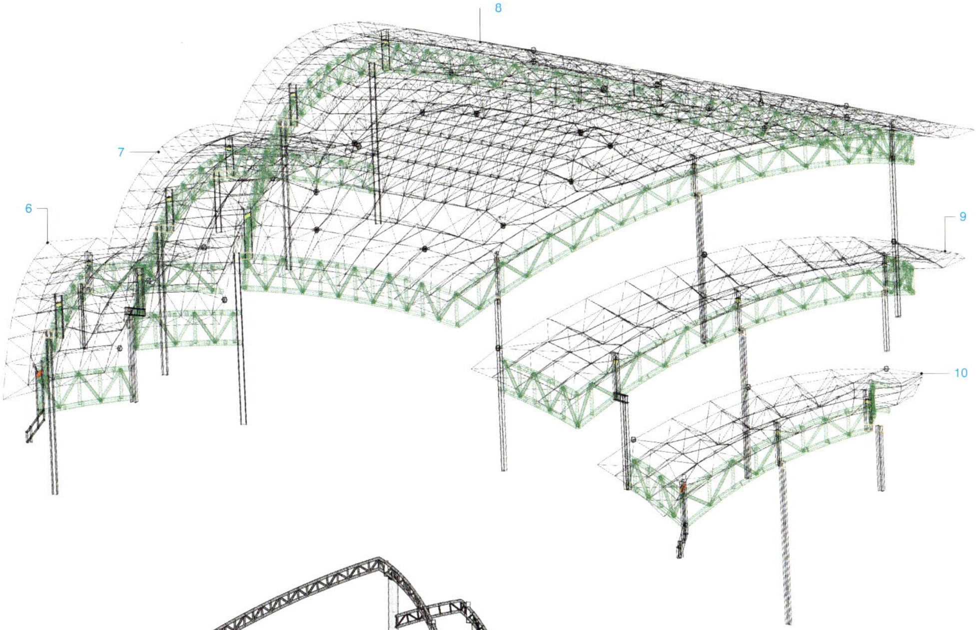

STEEL STRUCTURE AXONOMETRIC

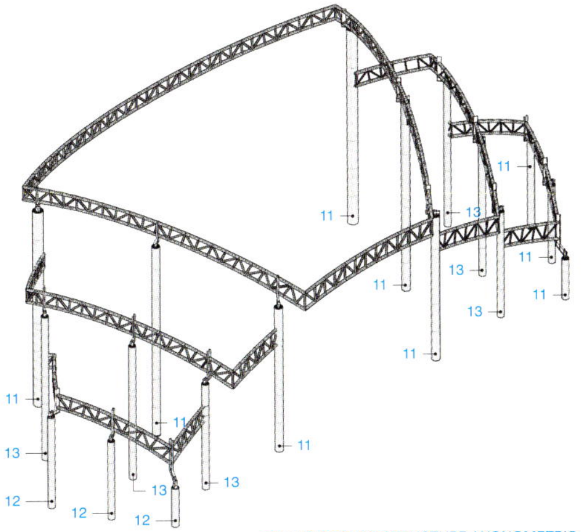

PILLAR & TRUSS STRUCTURE AXONOMETRIC

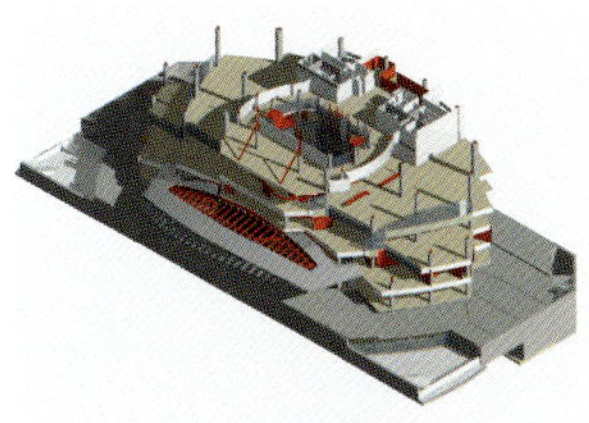

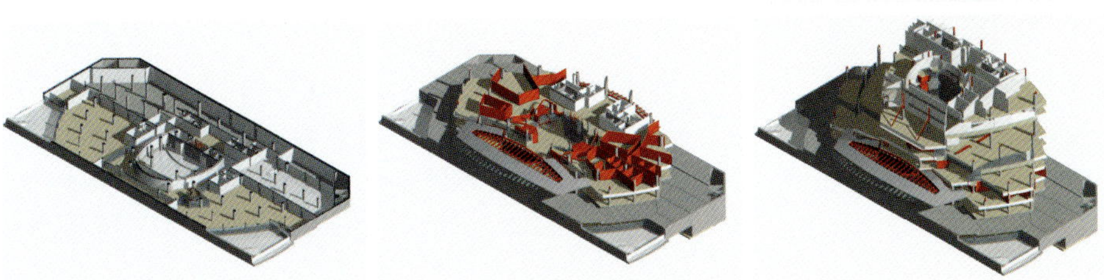

PROCESS

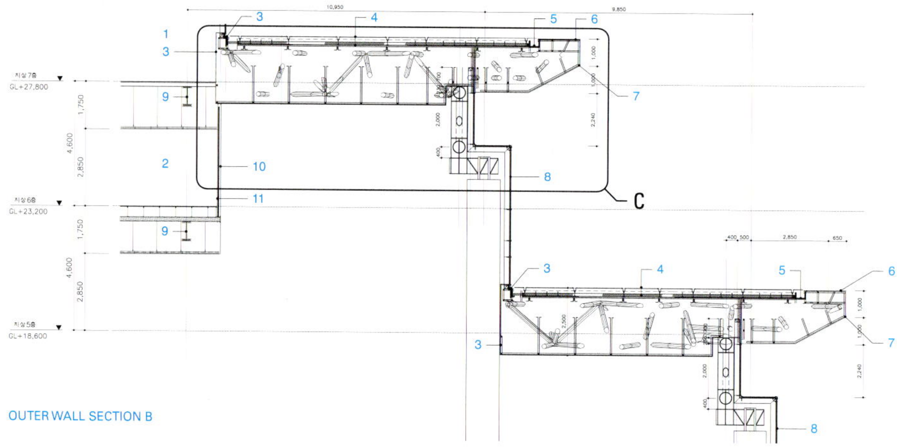

OUTER WALL SECTION B

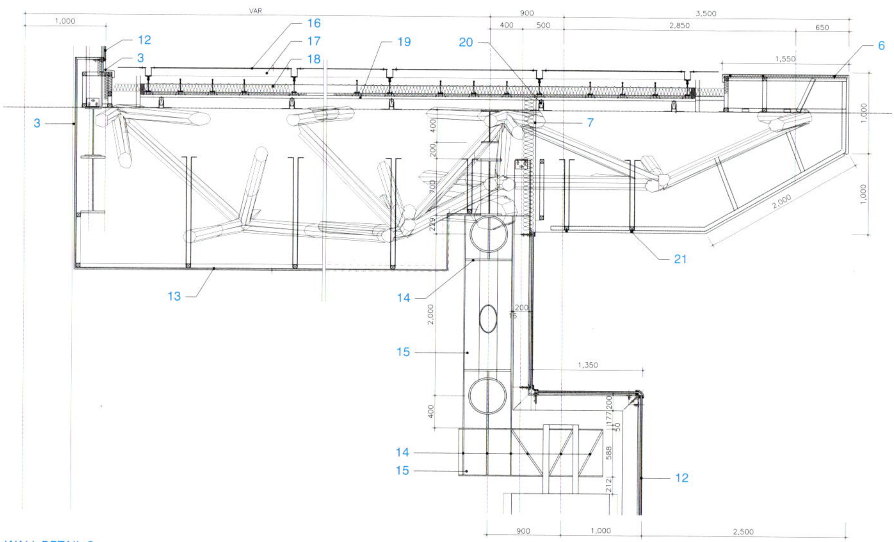

WALL DETAIL C

1 OFFICE
2 PROJECT ROOM
3 T1.2 STL PLATE / FLUOROCARBON RESIN PAIN-TING
4 PUNCHING PANEL / T125 GLASS WOOL INSUL-ATION BOARD
5 300X200 GUTTER
6 T3 AL. SHEET / □ - 50X50X3.2 STL PIPE
7 T125 ROCK WOOL PROTECTIVE BOARD / T1.2 STL PLATE / FLUOROCARBON RESIN PAINTING
8 T24 PAIRED, ONE SIDE SEMI-TEMPERED GLASS
9 T30 FIREPROOF COVER
10 T12.5 FIREPROOF GYPSUM BOARD 2 LAYERS / INTERIOR FIN
11 T50 MINERAL WOOL
12 T24 COLOR PAIRED GLASS
13 LIGHT GAUGE STEEL CEILING RIB / T9.5 GYPSUM BOARD / INTERIOR FIN
14 T20 STL PLATE
15 RUST-PREVENTING PAINT / FIREPROOF PAINT
16 COMPOSITE PANEL / T4 P.V.D.V. / APP. COLOR FIN
17 E.G.I.Z - BAR / 35X88X70XT1.2
18 #1/48K/T100 / INSULATION GLASS WOOL / INSULATION SKY VIVA 12/T30/ 88K
19 T1 E.G.I SHEET MILL FIN. / PORFORATION 1/2 PATTERN /Ø5/12 PICH
20 □ - 100X50X3.2T
21 WEATHER SILICONE SEALANT ON BACKER ROD SPONGE / FIELD FILL, DOWCORNING

Atrium open throughout eight floors

Ceiling view in the atrium

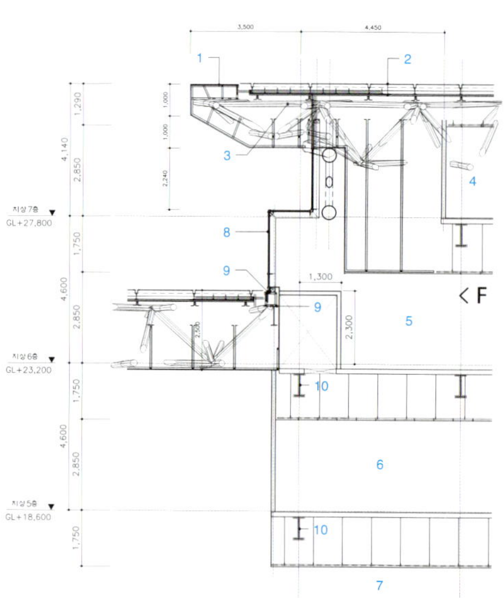

OUTER WALL SECTION D

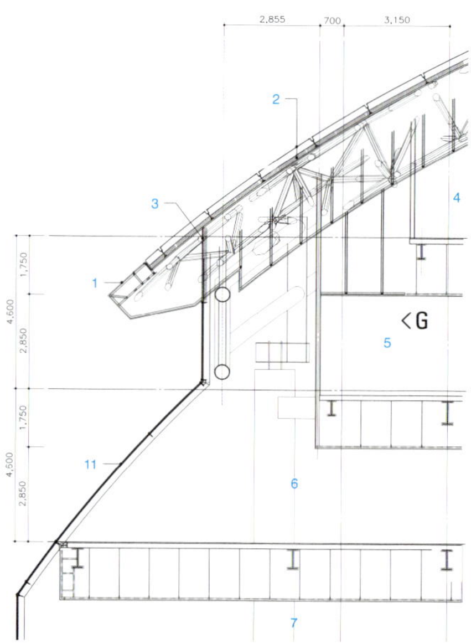

OUTER WALL SECTION E

1 T3 AL. SHEET / □- 50X50X3.2 STL PIPE
2 PUNCHING PANEL / T125 GLASS WOOL INSULATION BOARD
3 T125 ROCK WOOL PROTECTIVE BOARD / T1.2 STL PLATE / FLUOROCARBON RESIN PAINTING
4 PROFESSOR FACULTY ROOM
5 SERVER ROOM
6 FREE COMMUNITY ZONE
7 OPEN-STACK ROOM
8 T24 PAIRED, ONE SIDE SEMI-TEMPERED GLASS
9 T1.2 STL PLATE / FLUOROCARBON RESIN PAIN-TING
10 T30 FIREPROOF COVER
11 T30.76 LAMINATED, ONE SIDE SEMI-TEMPERED GLASS
12 H - 588X300X12X20
13 H - 596X199X10X15
14 H - 400X200X8X13
15 H - 700X300X13X24
16 PL - T24
17 PL - T20

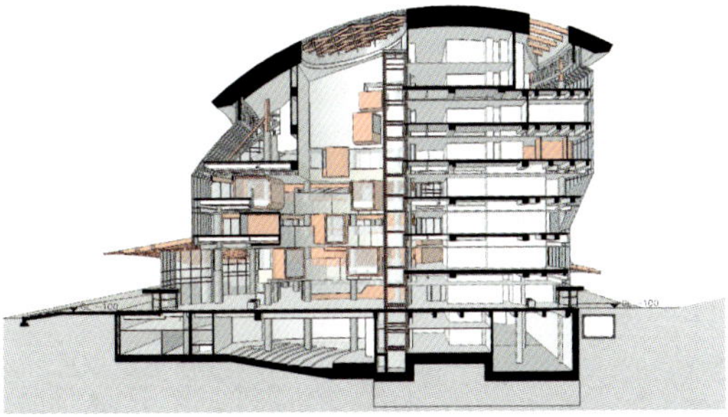

LONGITUDINAL SECTION

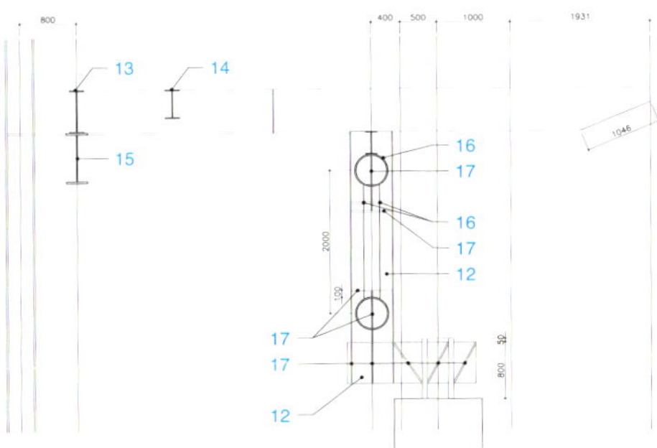

TRUSS & HAUNCH SECTION I

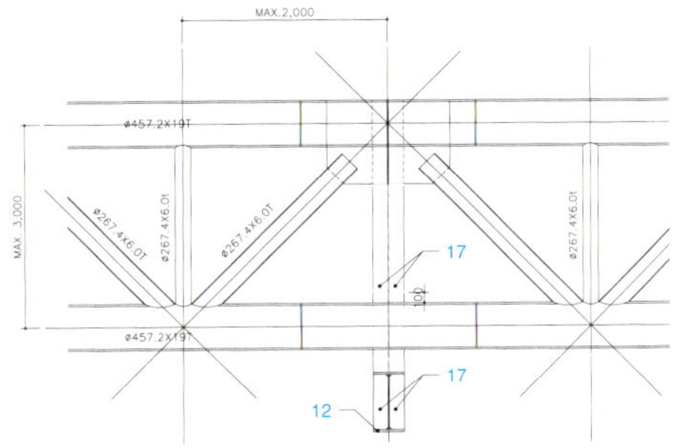

TRUSS & HAUNCH SECTION J

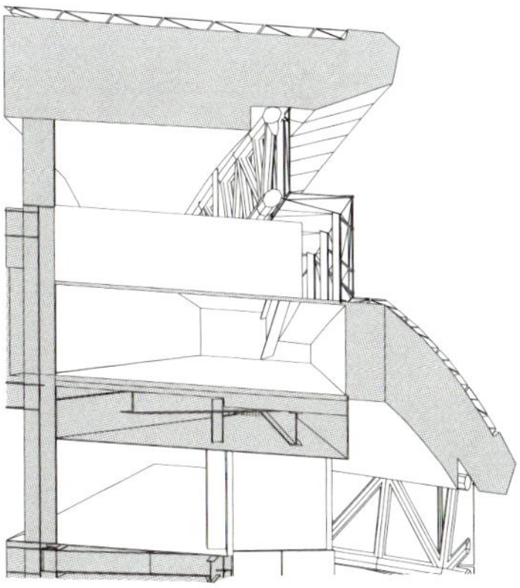

PERSPECTIVE F

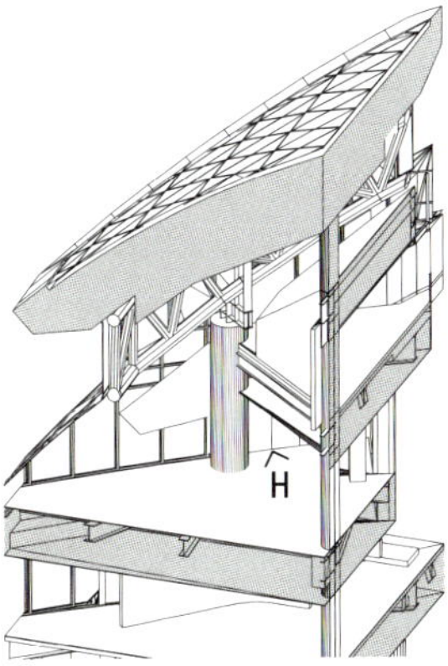

PERSPECTIVE G

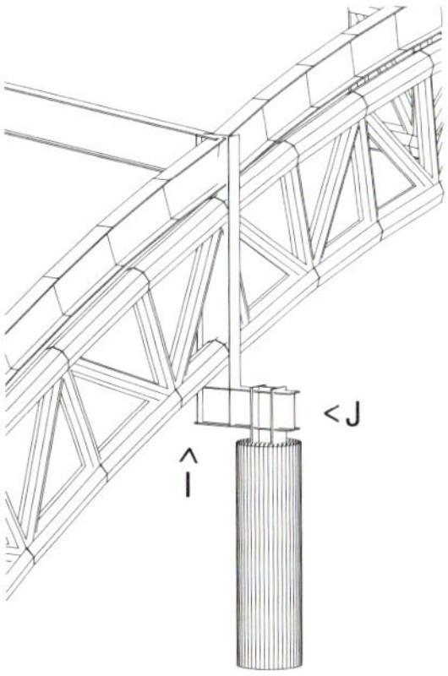

TRUSS & HAUNCH AXONOMETRIC H

Projecting masses of conference room

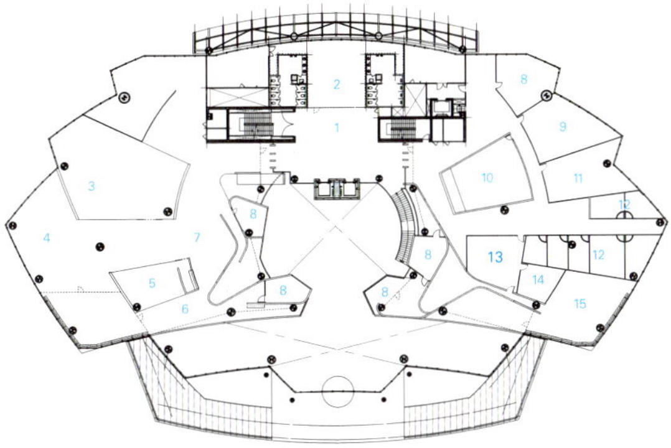

2ND FLOOR PLAN

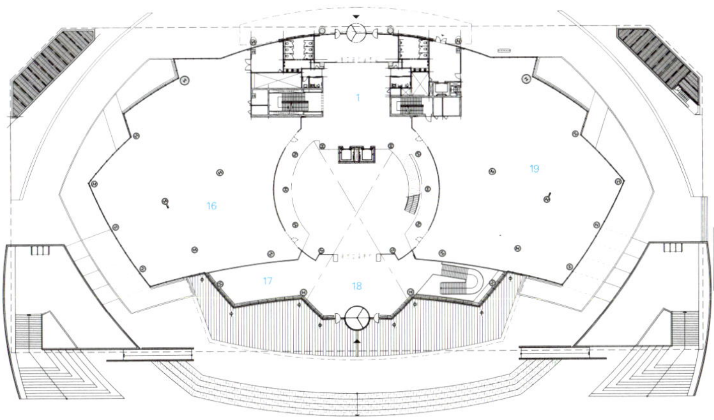

1ST FLOOR PLAN

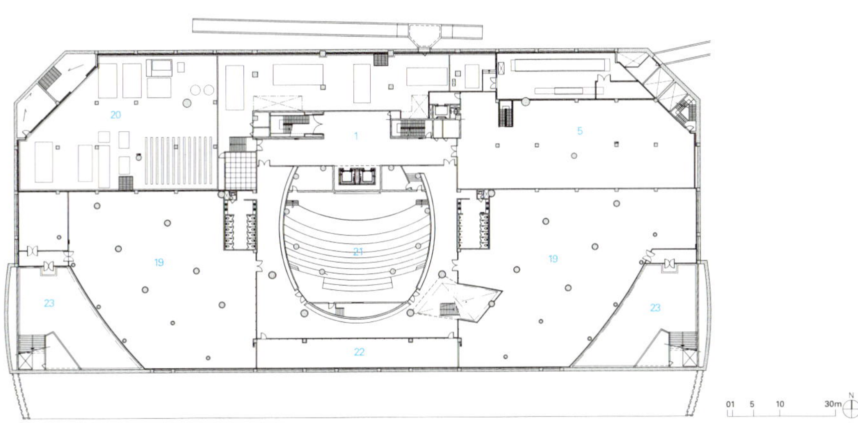

B1 FLOOR PLAN

1 HALL
2 REST AREA
3 VIDEO AREA
4 MULTIMEDIA AREA
5 LIBRARY
6 AUDIO AREA
7 READING ROOM
8 CONFERENCE ROOM
9 PRESENTATION ROOM
10 THEATER
11 SEMINAR ROOM
12 STUDY ROOM
13 CONTROL BOOTH
14 VIDEO BOOTH
15 MEDIUM PRACTICE ROOM
16 BROWSING ROOM
17 GIFT SHOP
18 LOBBY
19 FREE READING ROOM
20 MACHINE ROOM
21 AUDITORIUM
22 LOCKER
23 SUNKEN

Open stack reading room

Bookshelves and reading room

1 HALL
2 SYSTEM OPERATION
3 AIR CONDITIONING ROOM
4 CALL CENTER
5 SYSTEM DEVELOPMENT ROOM
6 PROJECT ROOM
7 PRINTER ROOM
8 SERVER ROOM
9 OPERATION ROOM
10 EXHIBITION
11 FREE COMMUNITY ZONE
12 REST ZONE
13 STAFF OFFICE
14 CONFERENCE ROOM
15 OPEN STACK READING ROOM

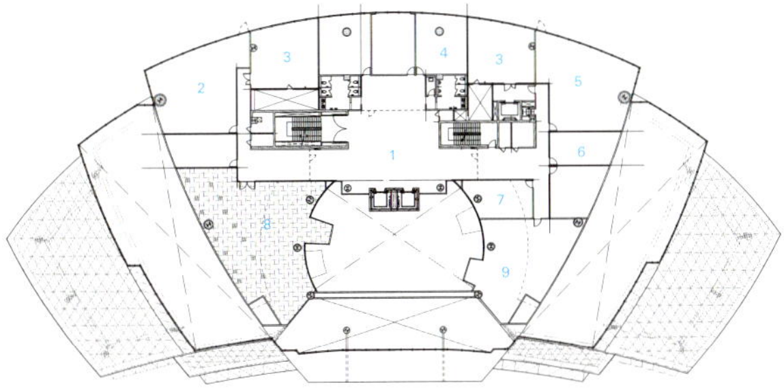

6TH FLOOR PLAN

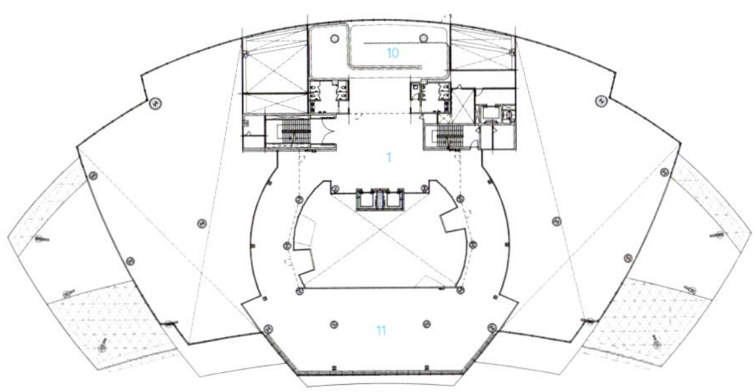

5TH FLOOR PLAN

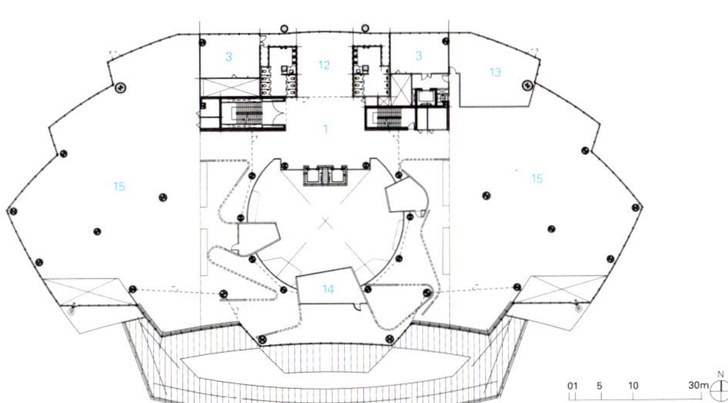

3RD FLOOR PLAN

Browsing room

Free community zone

Community zone and study room

Yellow conference room

Red staircase

KOREA UNIVERSITY HANA SQUARE

Location Anam-dong 5-ga, Seongbuk-gu, Seoul, Korea **Site Area** 640,377.00m² **Building Area** 717.49m² **Total Floor Area** 28,154.96m² **Building Scope** B3~1F **Structure** RC **Exterior Finish** Glass Point Fixing System(SPG), Aluminum Curtain Wall **Interior Finish** Granite, Stainless, Color Back Glass, Fabric **Parking** 526 Cars **Construction Period** 2005.1.11~2006.8.30 **Architecture Design** Koh Kwang-seog | SAMOO Architects & Engineers + Kang Myeong-ung | ACE Architects **Design Team** Kim Sang-joon **Construction** DOOSAN Engineering & Construction **Client** Korea University **Photographer** Lee Ki-hwan

The Hana Square of the Korea University project was planned as part of the Centennial Celebration Project of the Korea University. With the concept of the center of the campus, it is designed to function as the hub linking the four colleges of Science, Engineering, Life Sciences and Biotechnology, Information and Communication and the Library of Science. The Square houses an underground parking facility in order to address the problem of the whole campus turning into a car park because of the rapidly growing demand for parking space in recent years. Plants and gardens fill the open space on the ground from which the cars have been removed. The linear character of the site is emphasized by covering with glass the entire area of 200 by 50 meters. Along the length of the land, an atrium is created where the level height is higher than the ground, a sunken garden where it is lower. The Basement Levels 2 and 3 are occupied by a parking facility with a capacity of 520 cars in a total area of about 20,000 square meters. The Basement Level 1 accommodates diverse facilities essential for the students in about 9,200 square meters in three zones: student welfare, reading rooms, and mixed-use spaces. The student welfare zone consists of a bookstore, a convenience store and a food court. The reading room zone is filled with rooms for seminars and reading. The mixed-use zone comprises a multipurpose auditorium, an atrium, exhibition spaces and multi-media classrooms, all equipped with advanced technology. Natural light is filtered in through the atrium and the sunken garden, creating a bright and cheerful environment. As the Natural Science Campus is occupied mainly by the advanced research facilities of the information technology, science, engineering and biotechnology, the Hana Square is designed with modern and innovative appearance. The exposed glass curtain walls are built u sing a point fixing system, allowing the Square to function as a symbol reinforcing the high-tech image of the Natural Science Campus. As for Hana Square, we focused on the design to raise the symbolic identity through architectural object rather than creating a landscape architecture space on the ground. To prevent the object rising on the ground from being overwhelmed by the land, we selected glass as the surface of the object. The glass provides a clean and uncomplicated image by being simply slanted and the double skins for external surface are selected to stop simplicity from becoming boring. To do this, we adopted an external surface by using an SPG system for the insulating glass. By inserting the detailed interior in the simple mass from the outside, we pursue simplicity and complexity at the same time. To reduce heat gain caused by the external glass, we selected milk-white and transparent glass. We used insulated low-E glass charged with argon gas to achieve heat gain with safety. Between the double skins, we put a roll screen system to prevent solar heat to minimize a temperature rise from radiant heat. Moreover, we introduced the ventilation system to automatically control the temperature when it rises higher than a certain level. The system automatically senses the temperature rise and operates the windows to be open or closed. The base rock consisted of massive rock mass so the most important thing was to reduce the dug soil to minimize construction expenses. For the underground floor, if the height is too low, it would make the underground floor dark. It was important to keep the ceiling height to 3,200 millimeters and reduce the depth of the building. For the reason, the flat slab system was adopted and we reduced the thickness of the building in whole sections to minimize the dug amount. Located in the heart of the advanced research facilities, the Hana Square contributes to the symbolic identity of the campus. Making efficient use of space by applying an efficient parking system to create spaces for the pedestrians, it is also setting a new standard for environment-friendly space development. Text by Kim Sang-jun

General view from the north-east

1 ENTRANCE
2 LANDSCAPE
3 PARKING LOT
4 LIBRARY

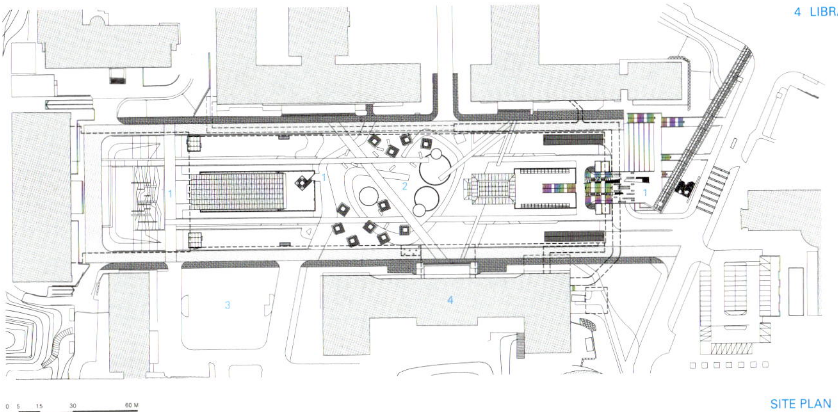

SITE PLAN

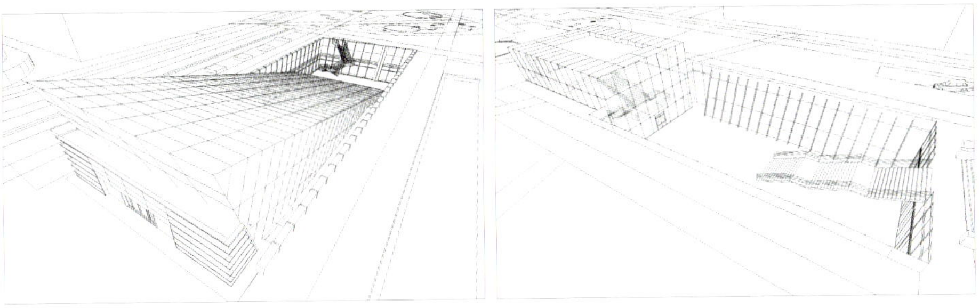

3D MODELING

Night view of architectural object

1 ELEVATOR HALL
2 ATRIUM(CONVENTION HALL)
3 SUNKEN GARDEN
4 PARKING LOT
5 CORRIDOR
6 MACHINE ROOM
7 HVAC ROOM
8 FAN ROOM
9 READING ROOM
10 GENERATOR ROOM
11 ELECTRIC ROOM

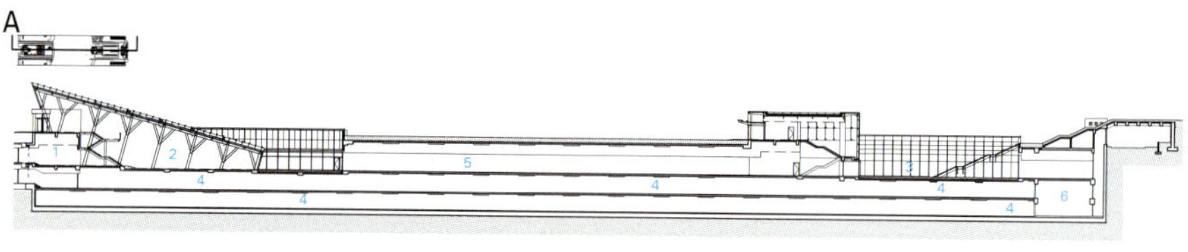

CROSS SECTION A

The glass mass of the exterior

The atrium viewed from the ground

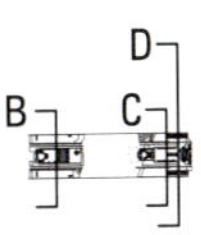

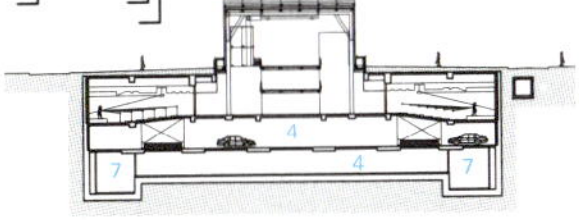

LONGITUDINAL SECTION B

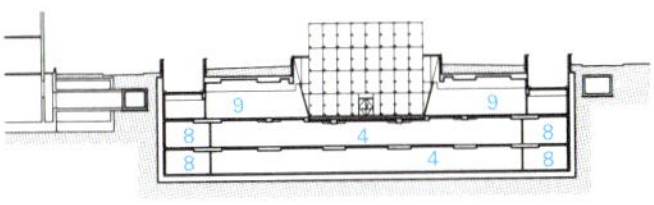

LONGITUDINAL SECTION C

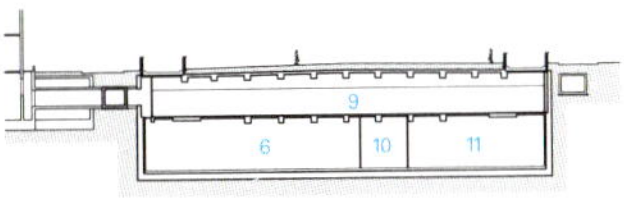

LONGITUDINAL SECTION D

The atrium(convention hall) on the b1 floor

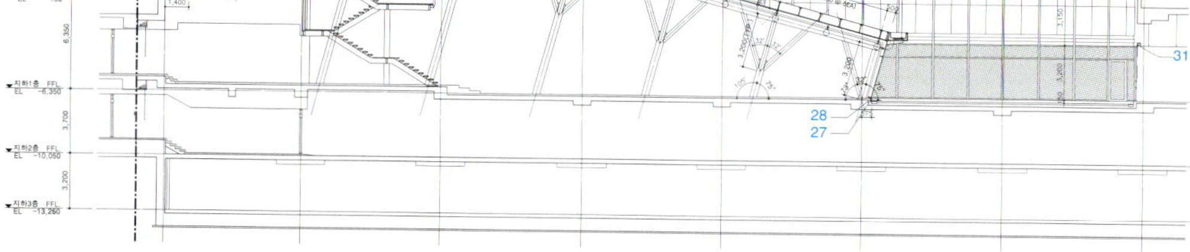

ARTRIUM SECTION

Pedestrian access is connected to the back sunken garden in from the entrance.

1 T3 ALUM. SHEET
2 12MM GLASS
3 4.5T BENT STEEL REINFORCEMENT
4 40MM ROUND BAR
5 M6 SCREW
6 M16 BOLT
7 T29.52 PAIRED LAMINATED GLASS
8 WEATHER SEALANT / W/BACK-UP ROD
9 STRUCTURAL SEALAN / W / NORTON TAPE12. 9T BENT STEEL PLATE
10 M8-BOLT & NUT
11 M12-BOLT & NUT
12 9T BENT STEEL PLATE
13 6T BENT STEEL PLATE
14 24MM GLASS
15 M10-BOLT & NUT
16 162.5MM STEEL PIPE
17 OPERATOR
18 5T ALUM ANGLE
19 ELECTRIC-POWERED ROLL BLIND
20 VENT
21 VENT FRAME
22 T24 WHITE PAIRED GLASS
23 T2 S'STL PANE
24 MANUAL REVOLVING DOOR
25 T29.52 WHITE PAIRED LAMINATED GLASS
26 T24 COLOR LOW-E PAIRED GLASS
27 ELECTRIC-POWERED INSPECTION WINDOW
28 T30 GRANITE FLAMING
29 T20 GRANITE FLAMING
30 GUTTER : T2 SST'L SHEET+ Ø200 SST'L LINE GUTTER
31 CURTAIN BOX /MANUAL ROLL BLIND
32 ⊏ -80X50X2.3T / BENT STEEL REINFORCEMENT
33 GALV. STEEL TUBE / LST-50X50X3.2T
34 T1.2 SST'L COVER
35 T2.0 SST'L SHEET / W/50MM INSULATION
36 T2.0 SST'L PLATE / MIRROR FINISH

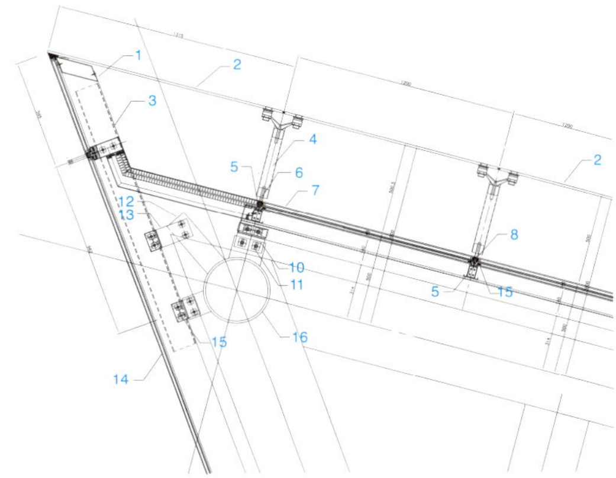

GLASS CEILING DETAIL E

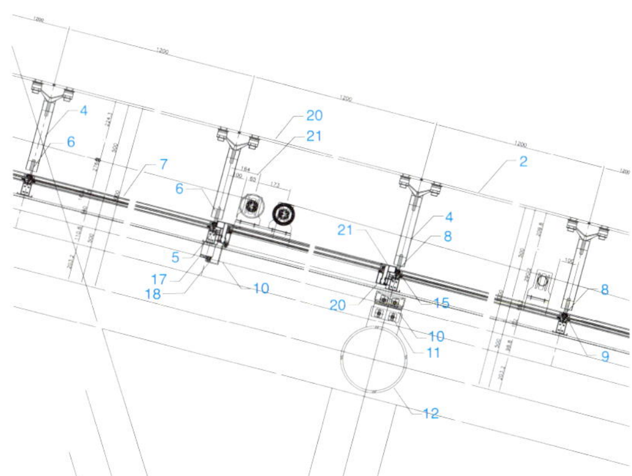

GLASS CEILING DETAIL F

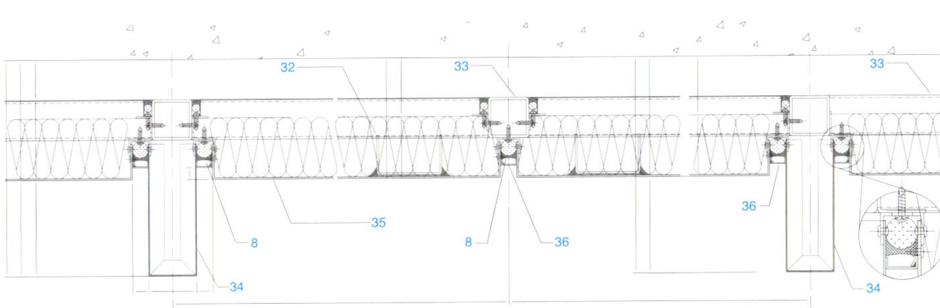

ATRIUM CEILING SECTION G

Sunken garden

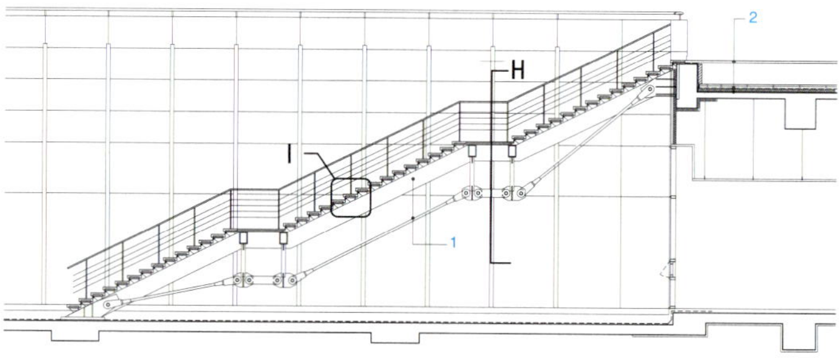

SUNKEN GARDEN STAIR ELEVATION

1 FLUOROCARBON RESIN PAINTING (METALIC SILVER)
2 REFER TO LANDSCAPING FIN. / MIN. T120 PLAIN CON'C MECHANICAL FIN. (W/WIREMESH #8-150X150) / T0.03 P.E FILM 2 FOLDS / T75 EXTRUDE METHOD INSULATING BOARD / T3 RUBBER ASPHALT SURFACE COATED WATERPROOFING
3 SST'L CAP
4 T70 GRANITE BOX STONE PROCESSING FLAMING
5 T10.76 LAMINATED GLASS (W/WHITE FILM)
6 SST'L ELLIPSE TYPE PROCESSED HAIR LINE FIN. (W=80
7 F.B-6X40 SST'L HAIR LINE FIN
8 2 - F.B-6X75 SST'L HAIR LINE FIN.
9 Ø6 SST'L WIRE
10 T30 GRANITE FLAMING / T30 CEMENT MORTAR / METAL LATH #800 WELDING ON STEEL PLATE / T20 STEEL PLATE
11 T10 STEEL PLATE / FLUOROCARBON RESIN PAINTING (METALIC SILVER)
12 T6 STEEL PLATE/ FLUOROCARBON RESIN PAINTING (METALIC SILVER)
13 □ - 400X200X20X20T/ FLUOROCARBON RESIN PAINTING (METALIC SILVER)
14 Ø101.6X5T/ FLUOROCARBON RESIN PAINTING (METALIC SILVER)
15 □ - 300X150X10X15T FLUOROCARBON RESIN PAINTING (METALIC SILVER)
16 Ø60 HOLE PROCESSING
17 T70 GRANITE BOX STONE PROCESSING FLAMING / T30 CEMENT MORTAR / METAL LATH #800 WELDING ON STEEL PLATE / T20 STEEL PLATE
18 SST'L SPIDER BRACKET
19 □ - 400X200X20X20T RUST-PREVENTING PAINT/ FIREPROOF PAINT/ FLUOROCARBON RESIN PAINTING (METALIC SILVER) (TYP.)

General view of the sunken garden from the interior

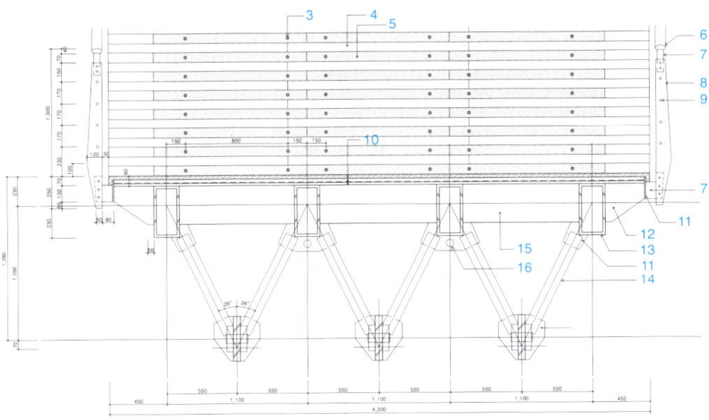

SUNKEN GARDEN STAIR SECTION H

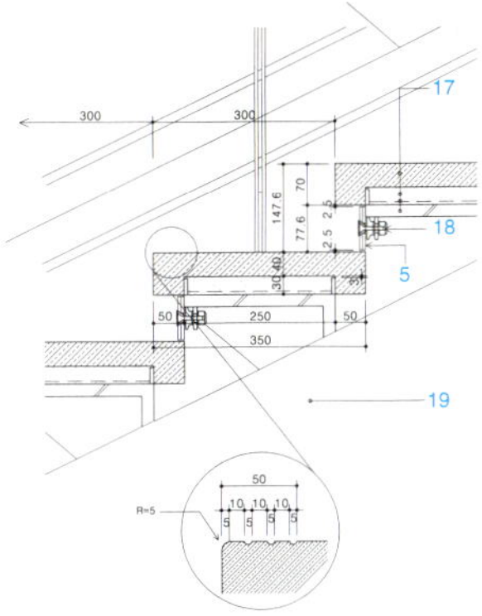

STAIR DETAIL I

A stair to the sunken garden

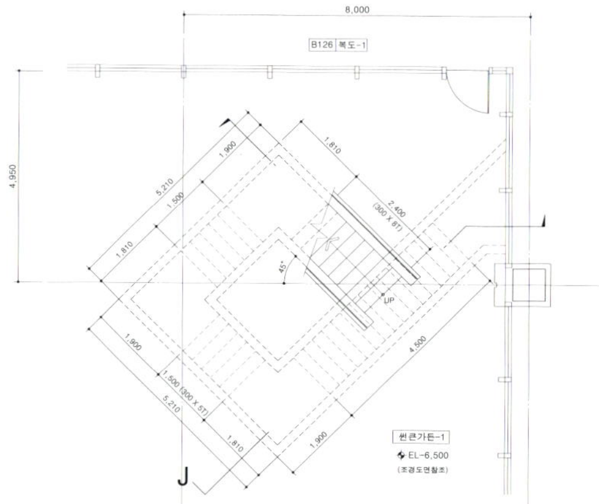

STAIR PLAN

1. ⌐-80X50X2.3T/BENT STEEL REINFORCEMENT
2. GALV. STEEL TUBE/LST-50X50X3.2T
3. WEATHER SELAN / W/BACK-UP ROD
4. 1.2T SST'L COVER
5. 2.0T SST'L SHEE / W/50MM INSULATION
6. 2.0T SST'L PLATE/MIRROR FINISH
7. □-800X200X35X40T/ METALIC SILVER FLUO-ROCARBON RESIN PAINTING
8. SST'L CAP
9. T80 GRANITE BOX STONE PROCESSING
10. T10.76 LAMINATED GLASS (W/WHITE FILM)
11. SST'L ELLIPSE TYPE PROCESSED HAIR LINE FIN. (W=80)
12. F.B-6X40 SST'L HAIR LINE FIN.
13. 2 - F.B-6X75 SST'L HAIR LINE FIN.
14. Ø15 SST'L PIPE MIRROR FIN.
15. T30 GRANITE FLAMING
16. T54 CEMENT MORTAR (W/METAL LATH #800 WELDING ON STEEL PLATE)
17. T6 STEEL PLATE
18. □-700X200X25X30T / METALIC SILVER FLUO ROCARBON RESIN PAINTING
19. □-500X150X10X15T / METALIC SILVER FLUO ROCARBON RESIN PAINTING
20. T80 GRANITE BOX STONE PROCESSING FLAMING / T24 CEMENT MORTAR / METAL LATH #800 WELDING ON STEEL PLATE / T6 STEEL PLATE
21. SST'L SPIDER BRACKET
22. T10.76 LAMINATED GLASS (W/WHITE FILM) (5+0.75+5)
23. METALIC SILVER FLUOROCARBON RESIN PA INTING

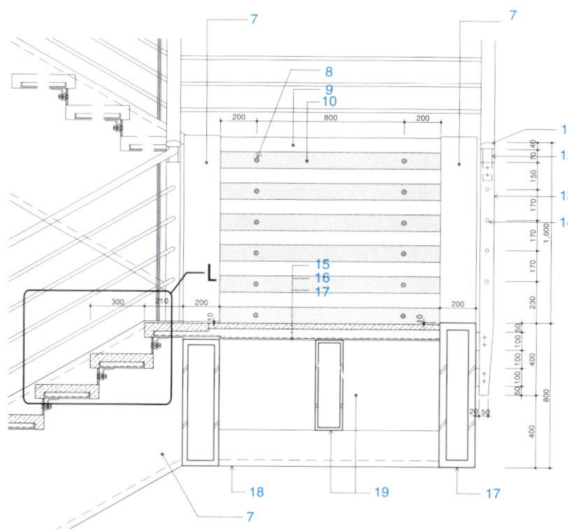

DETAIL K

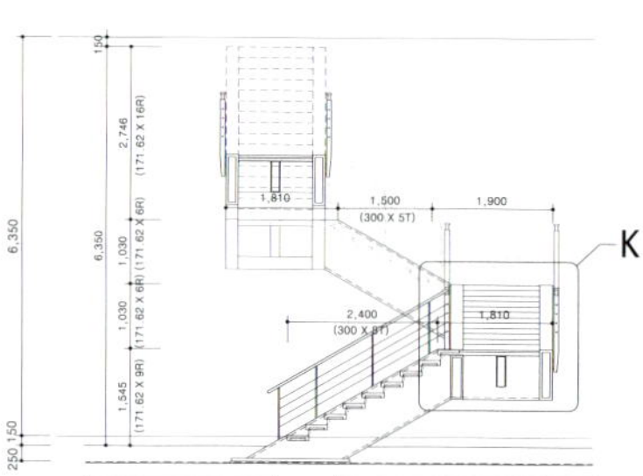

SECTION J

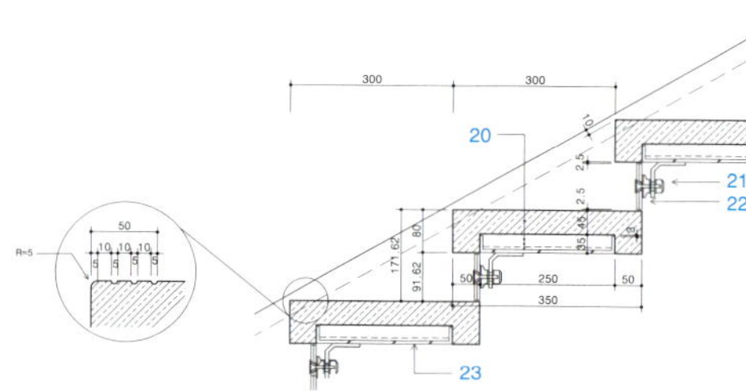

STAIR TRESTLE DETAIL L

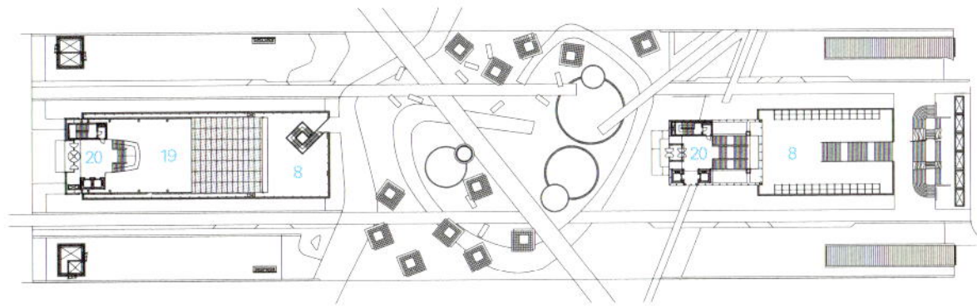

1ST FLOOR PLAN

B1 FLOOR PLAN

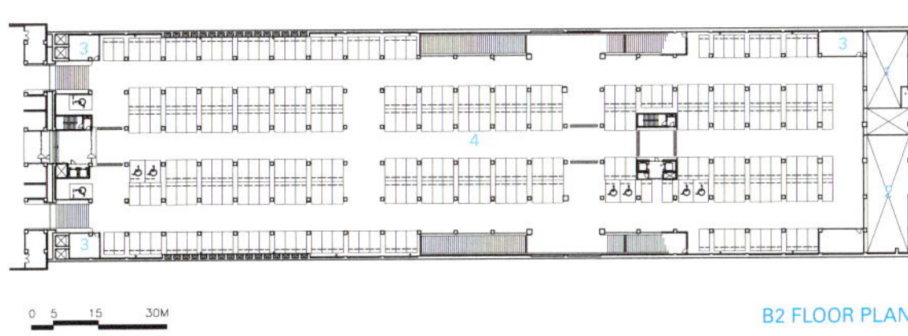

B2 FLOOR PLAN

1 UPPER ELECTRIC ROOM(OPEN)
2 UPPER MACHINE ROOM(OPEN)
3 FAN ROOM
4 PARKING LOT
5 LECTURE HALL
6 ATRIUM(CONVENTION HALL)
7 FITNESS CENTER
8 SUNKEN GARDEN
9 MULTIMEDIA ROOM
10 WELFARE FACILITIES FOR STUDENT
11 LIBRARY
12 DATE ROOM
13 INFORMATION
14 ELECTRIC ROOM
15 MACHINE ROOM
16 RESTING AREA
17 READING ROOM
18 LOUNGE
19 UPPER ATRIUM
20 LOBBY

The resting area for student

The corridor of the welfare facilities for student

>> EDUCATION

SEOKGWAN-DONG CAMPUS OF K-ARTS

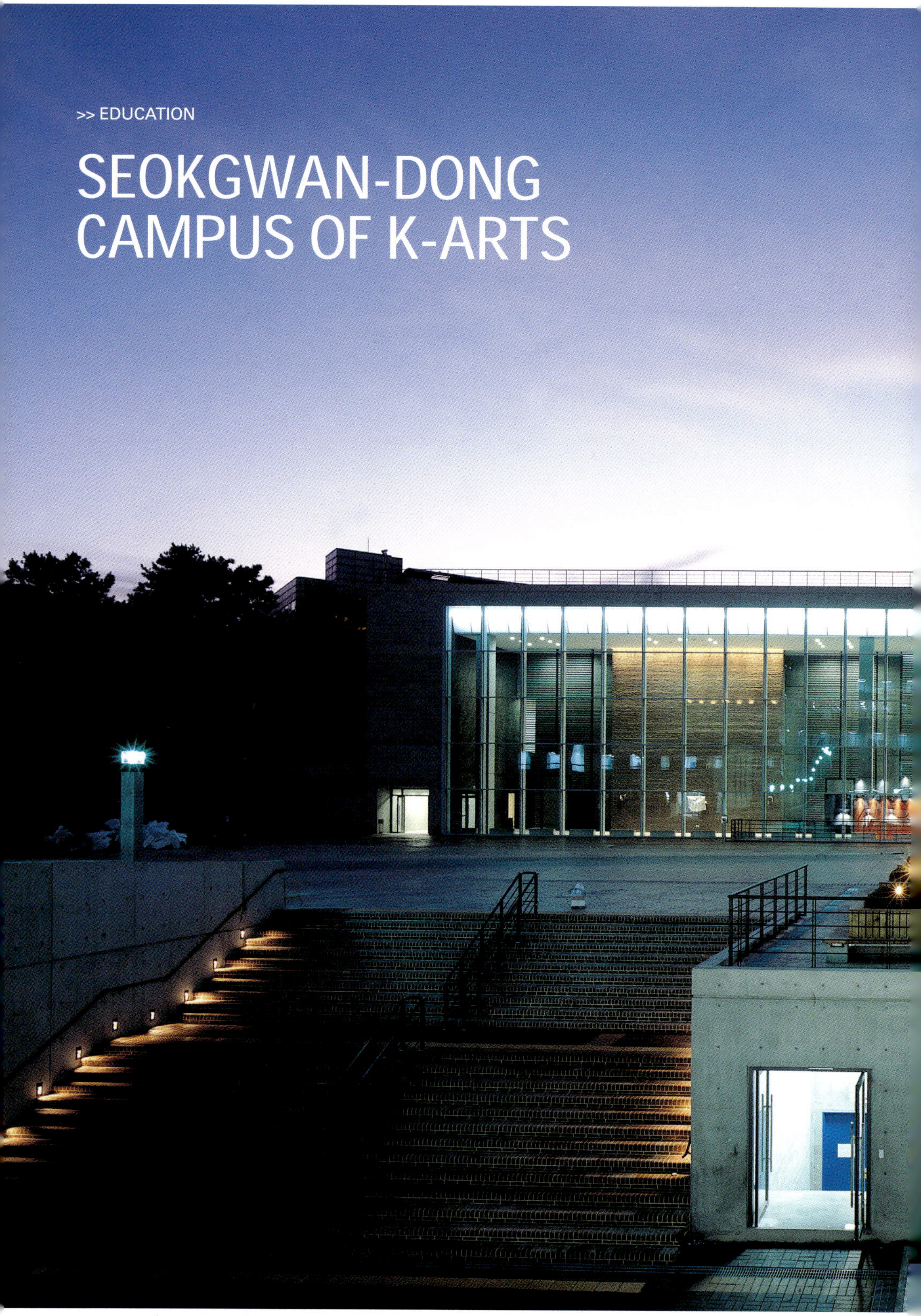

Location Seokgwan-dong, Seongbuk-gu, Seoul, Korea **Site Area** 433,520.00m² **Building Area** 14,517.63m² **Total Floor Area** 56,204.03m² **Landscape Area** 39,701.00m² **Building Coverage Ratio** 3.35% **Floor Area Ratio** 8.29% **Building Scope** B2~5F **Structure** RC, Steel **Exterior Finish** Exposed Concrete, T24 Pair Glass, Fluorocarbon Resin Corrugated Steel Plate, Split Paste Block, AL Louver **Interior Finish** Exposed Concrete, Squared Lauan, Clay Brick, Cement Block, Concrete Floor Tempered Material **Design Period** 2001~2002 **Construction Period** 2003~2006 **Architecture Design** Kim Kwan-joong | SAMOO Architects & Engineers + Lee Sang-leem, Oh Seom-hoon | SPACE GROUP **Design Team** SAMOO_Kim Hwan, Park Bong-gyu, Choi Seong-won, Lee Seon-hui, Park Sang-don / SPACE_Hong Sun-jae, Bae Yeon-su, Jeon Seong-tae, Kim Mi-su **Construction** SSANGYONG Engineering & Construction **Client** Korea National University of Arts **Photographer** Lee Ki-hwan

General view from the south

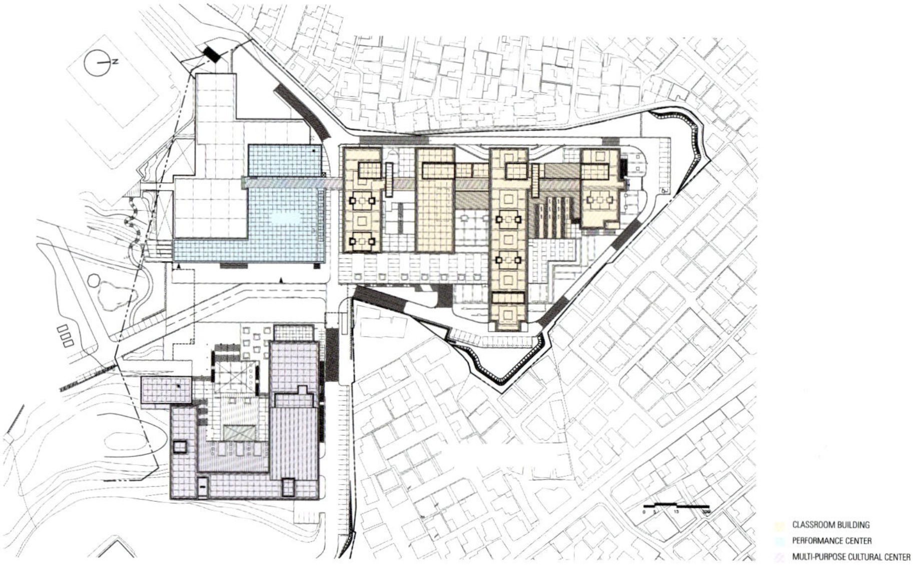

SITE PLAN

The Korea National University of Arts(K-Arts) focuses on the tutorage of experimental art and innovative design. The design of the new campus for K-Arts intends to inspire the insights of students and their creativeness. The exposed concrete surface on the buildings will become a background for their creative artworks.

The new campus, approximately 15km distant from the Seocho-dong existing campus, lack the cultural activities in its vicinity. To promote activities in the neighborhood, new campus plan intends to improve the public accessibility and convenience. Besides, buildings kept limited ceiling heights to conform to the architectural context of the neighborhood that is designated as a significant cultural property by Cultural Property Protection Agency.

The campus is zoned in three components; multi-purpose cultural center, performance center, and classroom buildings. The spatial composition of the campus can be related to the school's curriculum of freshman, sophomore, junior, and senior. One can also analogically infer it to the scenario composition of introduction, development, turn, and conclusion.

The open space between the cultural center and the performance center is occupied only by pedestrians, from which the paths to the each building diverge. The various programs in the multi-purpose cultural center are connected by the ramps and decks, while the classroom buildings of A, B, C, and D are linked by the horizontal circulation spine and vertical cores.

The cultural center, containing administrative offices, library, and student center, folds its massing to a U-shape configuration and creates a courtyard where sunken platforms, ramps, and decks dynamically compose the space.

The new campus of Korea National University of Arts is expected to become a major cultural institution in the northeastern district of Seoul, just as Seoul Arts Center plays in the southern area of Seoul. It will encourage the local public to more engage in the cultural activities and contribute to the quality of urban life.

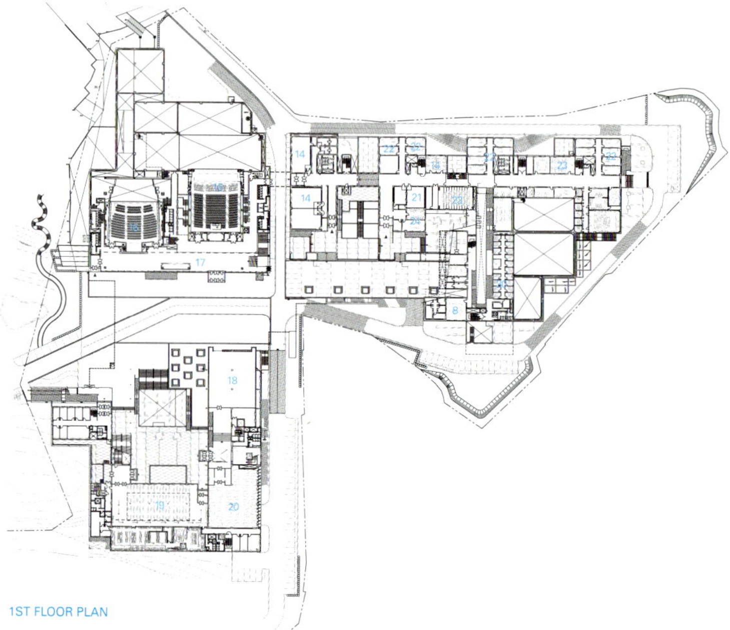

1ST FLOOR PLAN

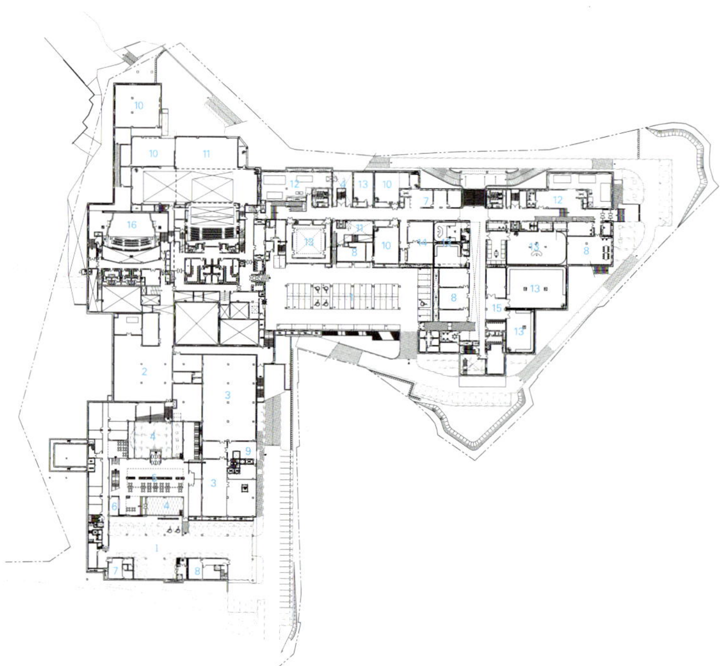

B1 FLOOR PLAN

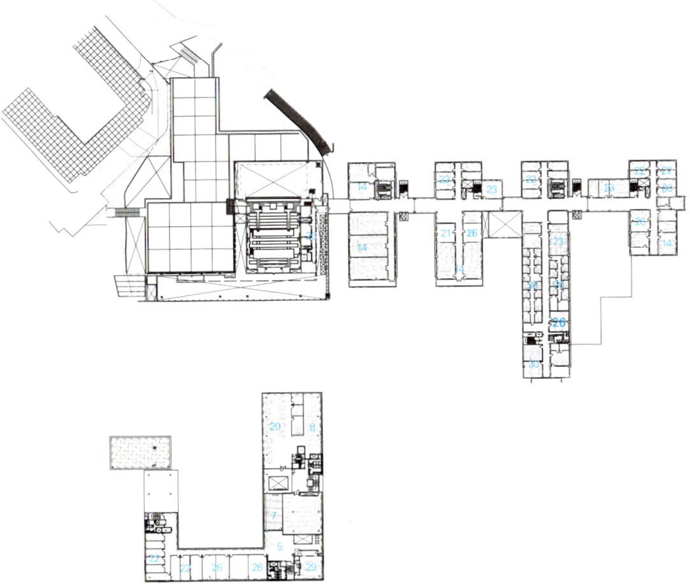

3RD FLOOR PLAN

1 PARKING LOT
2 ELECTRIC ROOM
3 LIBRARY
4 SUNKEN
5 HALL
6 SNACK BAR
7 RESTING ROOM
8 STORAGE
9 FILTRATION PLANT
10 MAKING ROOM
11 STAGE
12 HVAC ROOM
13 STUDIO
14 PRACTICAL ROOM
15 DRESSING ROOM
16 THEATER
17 LOBBY
18 EXHIBITION GALLERY
19 CAFETERIA
20 READING ROOM
21 STUDY ROOM
22 PROFESSOR'S OFFICE
23 LECTURE ROOM
24 SEMINAR ROOM
25 CIRCLE ROOM
26 OFFICE
27 OUTDOOR DECK
28 PROFESSOR'S LOUNGE
29 CONFERENCE ROOM
30 EDITORIAL ROOM

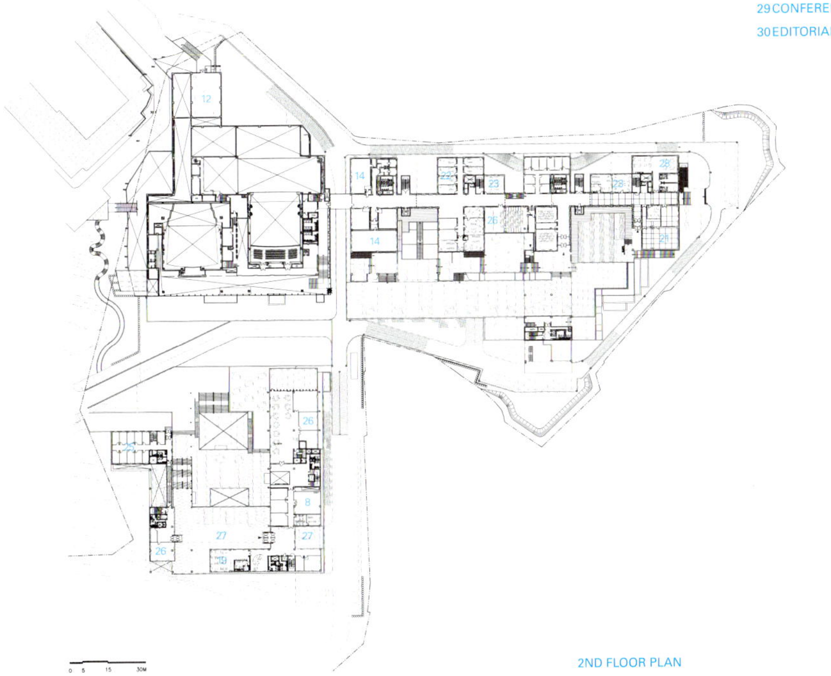

2ND FLOOR PLAN

Night view of the classroom building

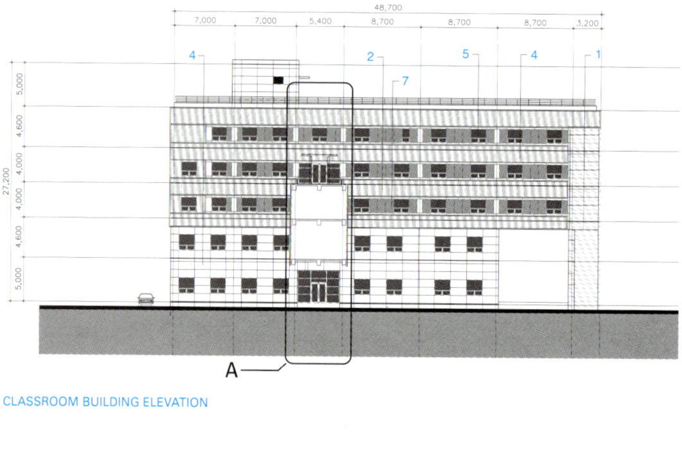

CLASSROOM BUILDING ELEVATION

1 ALUMINUM LOUVER
2 T0.8 FLUOROCARBON RESIN CORRUGATED STEEL PLATE
3 T2.3 FLUOROCARBON RESIN CORRUGATED STEEL PLATE (PERFORATED RATE 60%)
4 EXPOSED CON'C
5 T18 PAIRED GLASS
6 BLOCK FAIR-FACED LAYING
7 T0.8 FLUOROCARBON RESIN STEEL PLATE
8 V-115 T0.8 FLUOROCARBON RESIN STEEL PLATE
9 FLASHING T0.8 FLUOROCARBON RESIN STEEL PLATE
10 SILICON
11 BASE FLASHING
12 □-30×30×2.3T (BY OTHER'S)
13 T0.8 GALVANIZED CORRUGATED STEEL PLATE (FLAT PLATE)
14 T0.8 GALVANIZED CORRUGATED STEEL PLATE
15 T10 CLEAR TEMPERED GLASS
16 SPLIT PASTE BLOCK
17 CLOSURE / FLASHING / T0.8 FLUOROCARBON RESIN STEEL PLATE

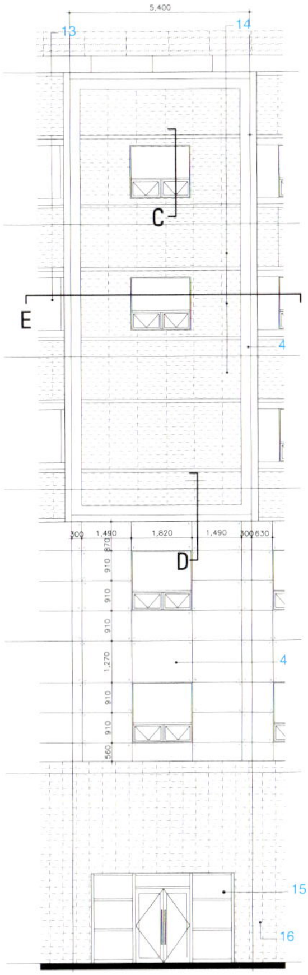

WALL DETAIL A

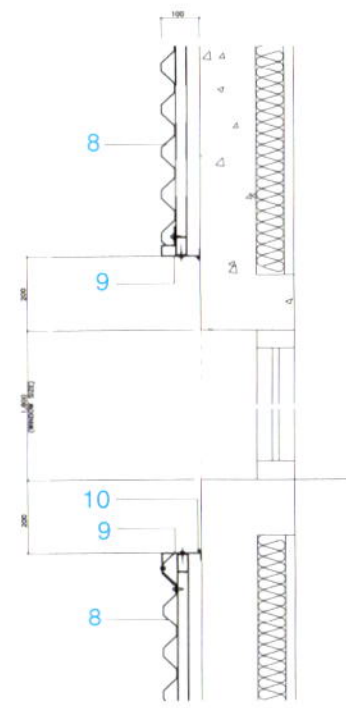

WINDOW SECTION C

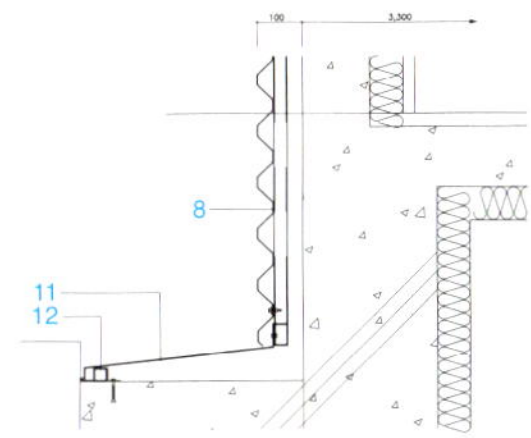

BASE SECTION D

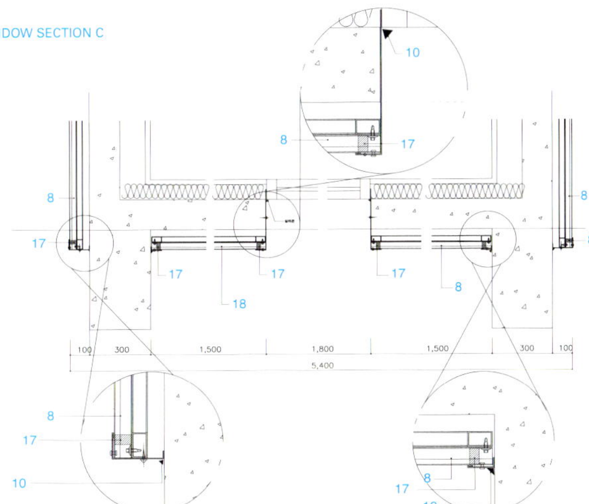

WALL WINDOW SECTION E

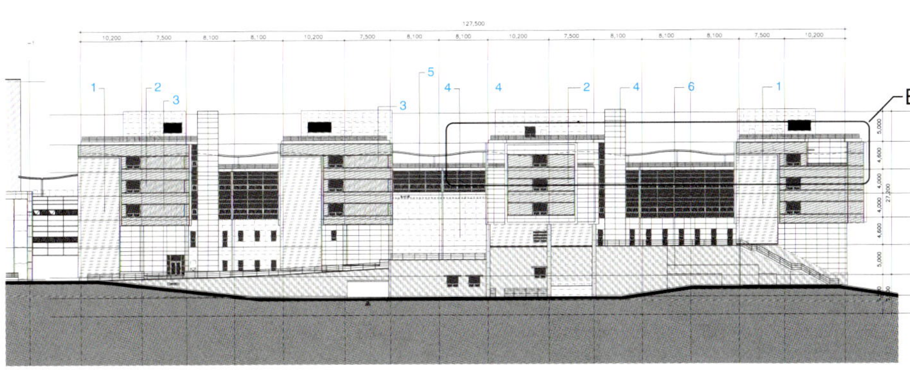

CLASSROOM BUILDING ELEVATION

The classroom building viewed from the east. The spaces are linked to one another by the spine.

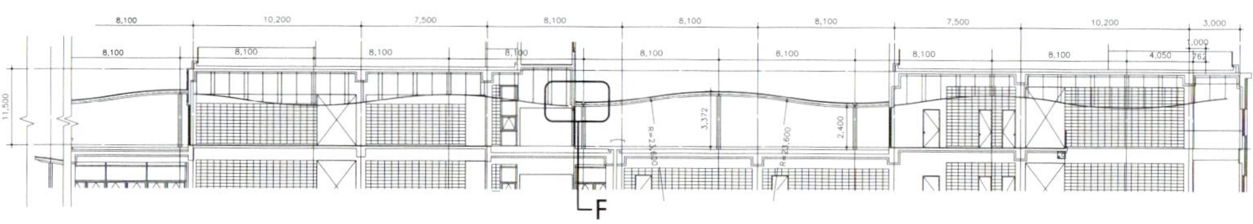

SPINE DETAIL B

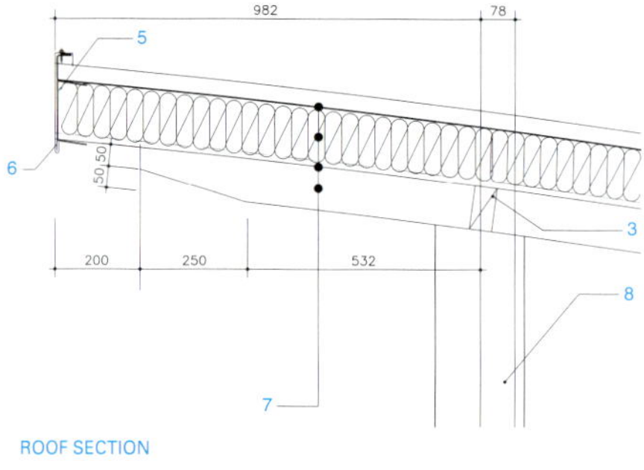

ROOF SECTION

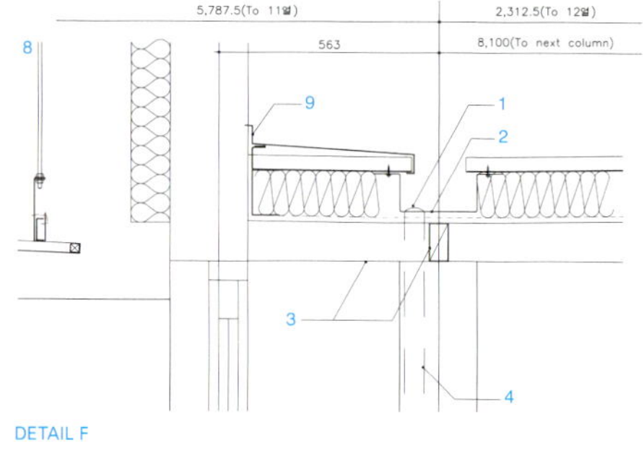

DETAIL F

General view of the classroom building from the south

1 Ø50 RAINWATER DRAIN
2 GUTTER W200 × H100
3 GALVANIZED SQ-PIPE 100×50×3.2T / FLUOROCARBON RESIN COATING / APP. COLOR
4 Ø50 RAINWATER EMBEDDING
5 END CAP
6 GALVANIZED FLASHING / FLUOROCARBON RESIN COATING / APP. COLOR
7 T0.6 GALVALUME ARCH PANEL / T120 ROCK WOOL BOARD (80K) / T0.6 GALVANIZED STEEL PLATE / GALVANIZED SQ-PIPE 100×50×3.2T / FLUOROCARBON RESIN COATING / APP. COLOR
8 SQ-PIPE 150×150×6T(SPSR400) / FLUORO-CARBON RESIN COATING / APP. COLOR
9 FLASHING

The multi-purpose cultural center viewed from the classroom building

General view from the multi-purpose cultural center

1 EXPOSED CON'C
2 WOODEN PANEL (RUSTPROOFING TREATMENT)
3 WOODEN LOUVER
4 LIGHT-WEIGHT ROOF PANEL

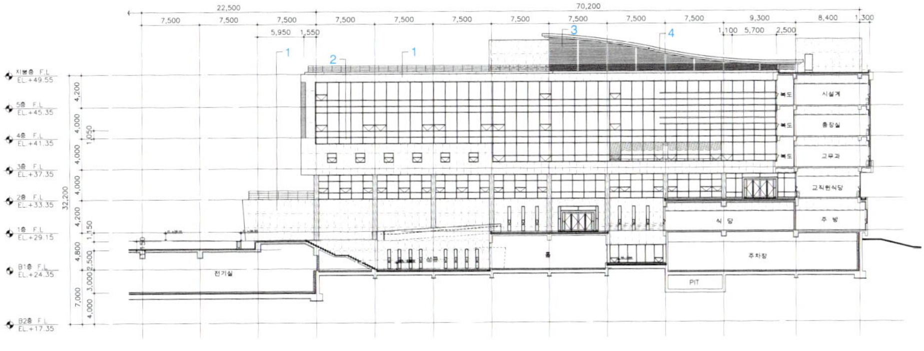

MULTIPURPOSE CULTURAL CENTER LONGITUDINAL SECTION

1 RESTING ROOM
2 PROFESSOR'S OFFICE
3 CORRIDOR
4 KITCHEN
5 CAFETERIA
6 PARKING LOT
7 SNACK BAR
8 MEDICAL OFFICE
9 CIRCLE ROOM(NEWSPAPER)
10 ELECTRIC ROOM
11 PROJECTION ROOM
12 HVAC ROOM
13 DATA ROOM
14 READING ROOM
15 HALL
16 INFORMATION RETRIEVAL ROOM
17 EXHIBITION GALLERY
18 LIBRARY
19 WATER-PURIFIER TANK
20 LAB ROOM
21 STUDIO
22 FAN ROOM

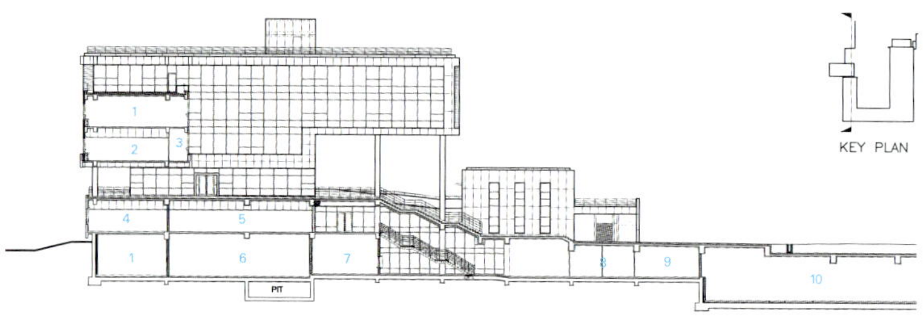

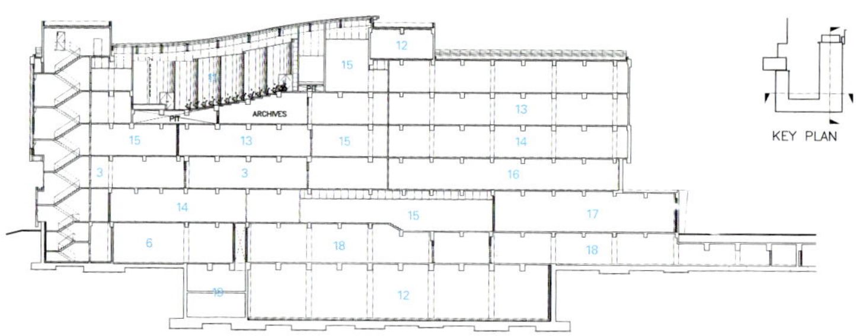

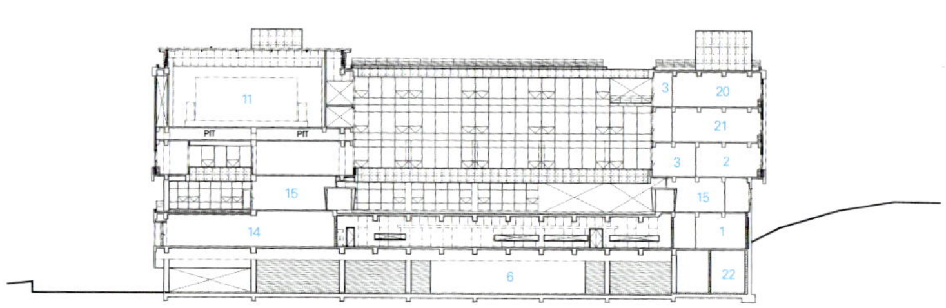

MULTIPURPOSE CULTURAL CENTER SECTION

Performance center viewed from the multipurpose cultural center

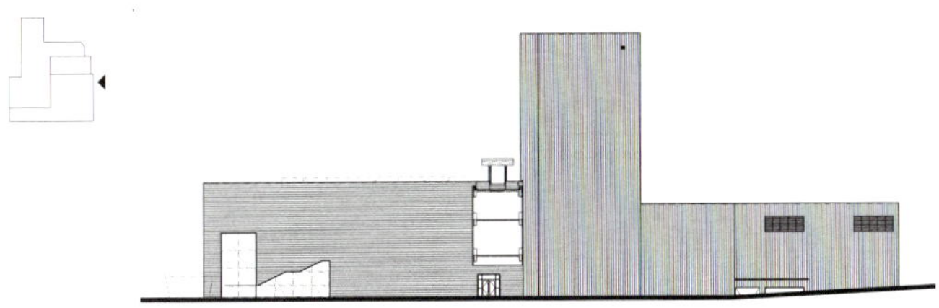

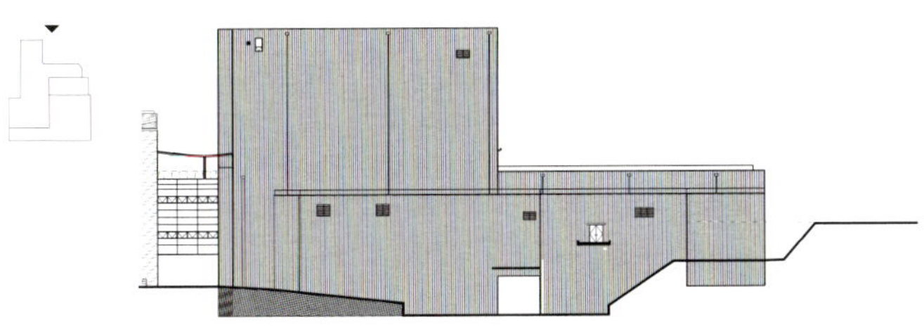

PERFORMANCE CENTER ELEVATION

The lobby of the performance center

1 CLAY BRICK 0.5B FACING MASONRY (INTERIOR WORK)
2 WOOD GRILL (INTERIOR WORK)
3 PAINT ON GYPSUM BOARD (INTERIOR WORK)
4 T9.5 GYPSUM BOARD 2PLY / APP. PAINT FIN.
5 FIXING HARDWARE
6 □150×100 T3.2 ST'L / 75×50 SOLID WOOD
7 FLOOR REINFORCING AGENT / CEMENT MORTAR / BUILDING SLAB
8 □ 51×1.6T SST'L (VIBRATION)
9 □19.1×1.2T STL PIPE
10 W50XT9 STL PL.
11 W25XL60×T12 STL PL.
12 GROUT GAP-FILLING
13 GALVA. ST PL-6T×180×180
14 30X30 REINFORCEMENT STEEL PIPE / T9 PLYWOOD 1PLY / T9.5 GYPSUM BOARD 1PLY / APP. PAINT FIN
15 APP. LIGHTING INSTALL (CF-1: SPECIFICATION REFERENCE)
16 STEEL FIXED WIRE
17 T9 FROST TEMPERED GLASS
18 T1.6 SST'L PL. (VIBRATION) / □ 40×40 (@900)
19 T120 SLOPED PLAIN CON'C (#8-100X150 WIRE MESH) / T27 PROTECTIVE MORTAR / MEMBRANE WATERPROOFING / STEM CON'C
20 HEAT INSULATION
21 CEILING : T18 PAIRED GLASS (ONE-FACE SEMI-TEMPERED GLASS)
22 T0.8 GALVANIZED CORRUGATED STEEL PLATE
23 MASONRY JOINT 10MM
24 ATTACHING MORTAR (10MM)
25 APP. BRICK 0.5B MASONRY
26 HANDRAIL TYPE 3 (STS)
27 HANDRAIL TYPE 4 (WOOD)
28 HANDRAIL TYPE 1 (WOOD)
29 HANDRAIL TYPE 1
30 HANDRAIL TYPE 3 (WOOD)

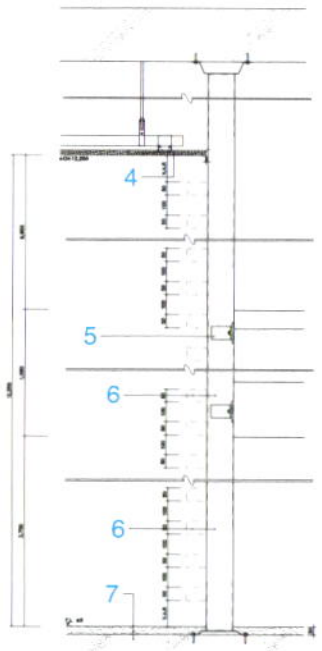

LOBBY WALL SECTION G

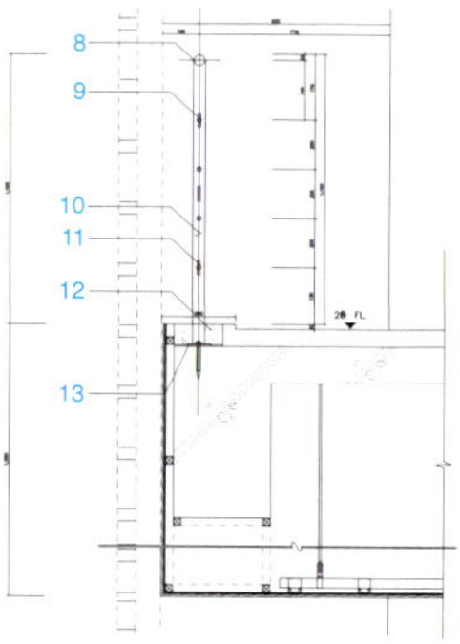

LOBBY HANDRAIL SECTION H

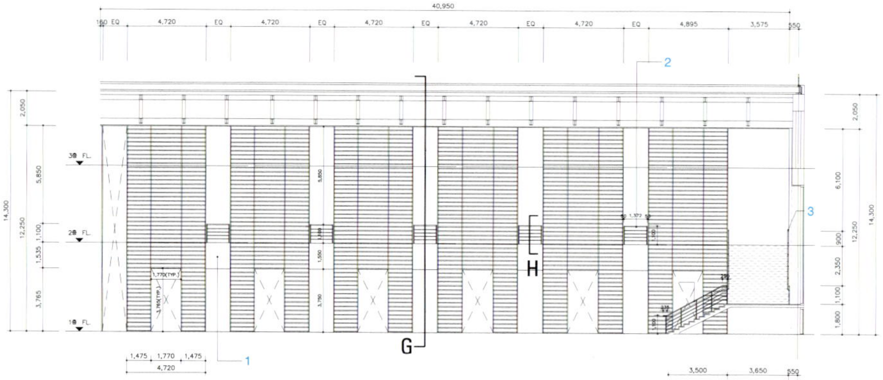

PERFORMANCE CENTER LOBBY ELEVATION

Lobby wall detail

Stair part

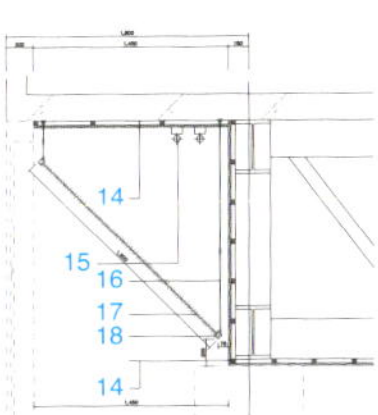

LOBBY CEILING SECTION I

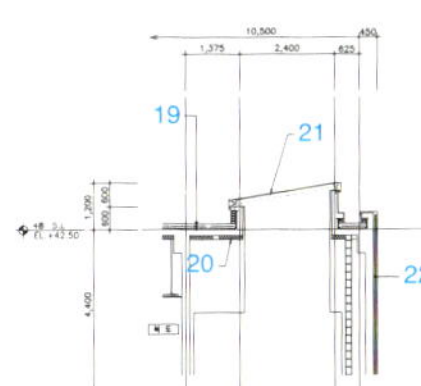

DETAIL J

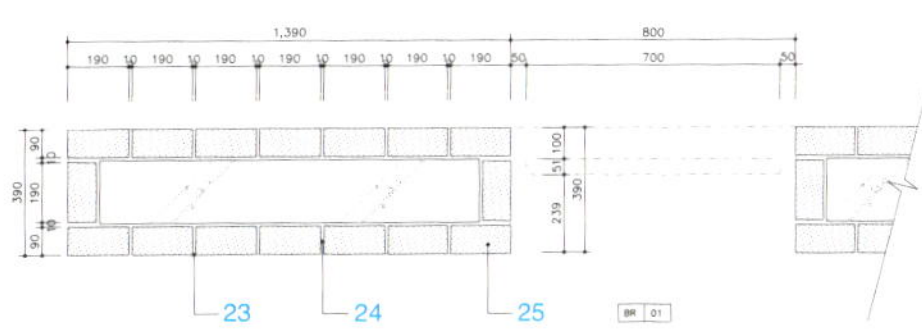

LOBBY WALL SECTION K

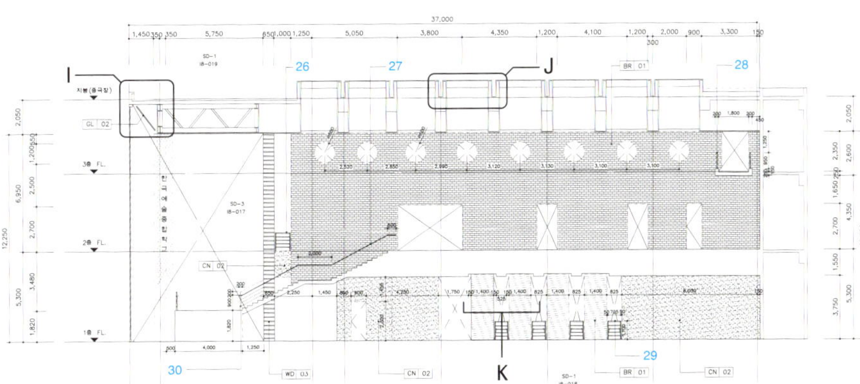

PERFORMANCE CENTER LOBBY ELEVATION

EWHA GIRLS' FOREIGN LANGUAGE HIGH SCHOOL VISION CENTER

Location Sunhwa-dong, Jung-gu, Seoul, Korea **Site Attribute** General Resident **Site Area** 54,125m² **Building Area** 10,090.22m² **Total Floor Area** 41,218.08m² **Building Coverage Ratio** 18.64% **Floor Area Ratio** 56.45% **Building Scope** B2~3F **Structure** Steel, RC **Exterior Finish** Aluminum Bar, IPE Wood, Color Pair Glass, Exposed Concrete **Design Period** 2004.5~2005.9 **Construction Period** 2005.9~2007.8 **Architecture Design** Kim Seung-hoy | Seoul National University + Kang Won-phil | KYWC Architects **Design Team** Jeong Hwa-taek, Park Se-min **Photograph** KYWC Architects

The site for the future vision center of the school has an irregular form and relatively steep slope. The existing Stranton center isolated from surroundings was asking for a change. The enlarged vision center was built in the consideration of properties of the site and the relation with existing buildings. Also a program with complex and various functions including a cafeteria, an auditorium, special classrooms and a teacher's room was demanded. The more the site gets complex and the more various demanded programs are, the simpler architectural concept is required. A simple system makes it possible to create something complicated. A uniformed - width band was set within the enlargement site with a triangle shape and named as 'program band'. The program is arranged in a lined space that has repetitive same modules according to required sizes. Program band reacts to a irregular form of land to constitute sequential space system that has continuation possibility in connection with existing buildings. Through program band, a yard was created in the center of the site. This yard named as 'program plaza' utilizes various levels of the site to hold the big indoor space and outdoor space. Each level, which is connected to the existing Stranton center, is expanded to the opposite side of the site to make a three dimensional space in a tiered form from the 1st basement to the 3rd above ground as a main space for students' activities. Program plaza that will be filled with Ewah students' various activities, such as, various performances and club activities, draws abundant architectural experiences as a main space of Ewha Girls' Foreign Language High School.

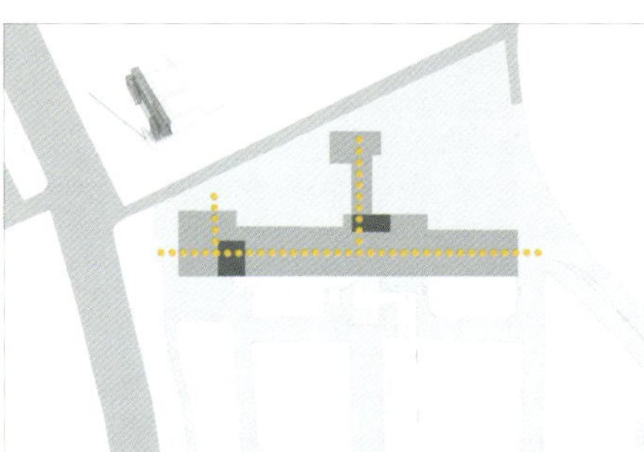

ISOLATED EXISTING BUILDING

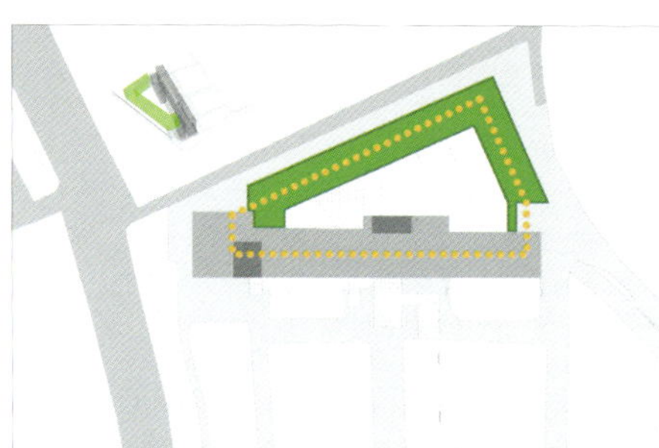

PROGRAM BAND: SUSTAINABILITY

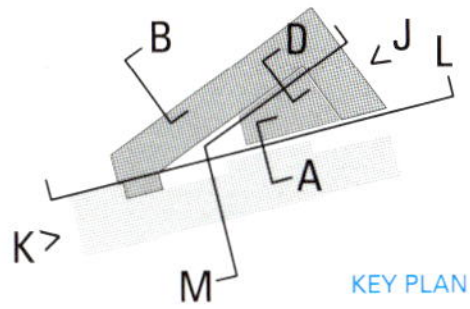

KEY PLAN

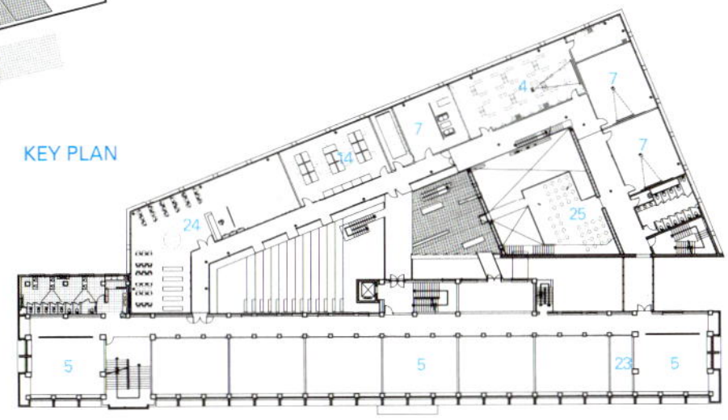

2ND FLOOR PLAN

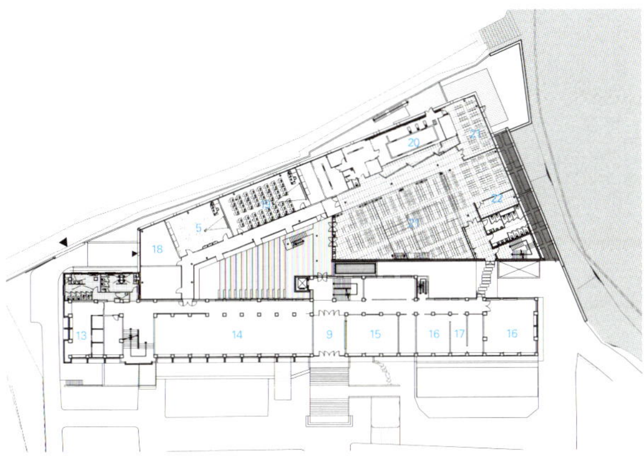

1ST FLOOR PLAN

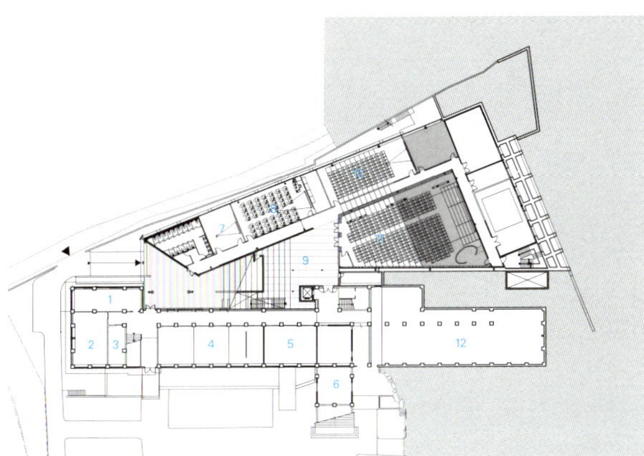

B1 FLOOR PLAN

1 PRINTING ROOM
2 INFIRMARY
3 COUNSELING ROOM
4 LABORATORY
5 CLASSROOM
6 STORAGE
7 FUNCTIONAL ROOM
8 LANGUAGE LAB
9 HALL
10 AV ROOM
11 MULTIPURPOSE ROOM
12 PIT
13 STUDIO
14 TEACHER'S ROOM
15 ADMINISTRATIVE OFFICE
16 CONFERENCE ROOM
17 PRINCIPAL'S OFFICE
18 RECESS ROOM
19 COMPUTER LAB
20 KITCHEN
21 CAFETERIA
22 SNACK BAR
23 EDITORIAL ROOM
24 LIBRARY
25 LOUNGE

Louver skin detail

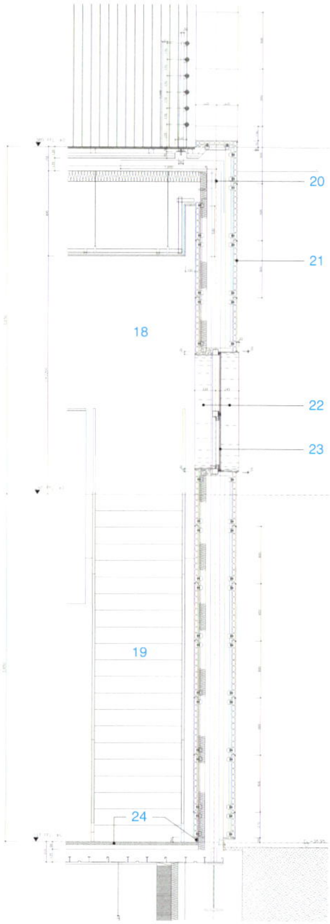

OUTER WALL SECTION A

1 SETTING T50 BEAN GRAVEL / INORGANIC AND ELASTIC SURFACE COATING WATER-PROOFING (EXPOSED TYPE)
2 PUT MORTAR
3 AL. FLASHING
4 T110 INSULATION(PIN UP STICKING)
5 T20 FIREPROOF SPRAYING(AN HOUR FIREPROOF)
6 CORNER REINFORCEMENT MDF
7 T1.6 GALVALUME, BANK, OIL PAINT(APPOINTED COLOR)
8 16MM CLEAR PAIRED GLASS + INSULATING FILM
9 AL. FLASHING 1
10 LIGHT GAUGE STEEL FRAME CEILING RIB(M-BAR) / VINYL PAINT AFTER ALL PUTTY-(APPOINTED COLOR) ON T9.5 GYPSUM BOARD TWOFOLD
11 GAUGE STEEL COLUMN : BANK, FIREPROOF, OIL PAINT(APPOINTED COLOR, FLATTING, 1 HOUR FIREPROOF)
12 CORRIDOR
13 WATER CUTING GUTTER
14 VINYL PAINT AFTER ALL PUTTY(APPOINTED COLOR) ON T9.5 WATERPROOFING GYPSUM BOARD TWOFOLD
15 T1.6 GALVALUME: BANK, OIL PAINT(APPOINTED COLOR)
16 DECO TILE(450X450)
17 T20 IPE WOOD
18 LOUNGE
19 STUDENTS RESTAURANT
20 Ø16@400 STUD BOLT(H=100)
21 T40 BASE PANEL(W=600, L=2,400)
22 T5 PL.: BANK, OIL PAINT(APPOINTED COLOR, FLATTING)
23 AL. CIRCLE WINDOW(LOWER PART PROJECT OPEN)
24 T30 GOHONG STONE FLOOR RUBBING (450X1,200)
25 FLOOR FIREPROOF PARTITION RECHARGING W/T1.6 GALVALUME
26 INSTALLING AL. GRILL
27 INSTALLING ROLL BLIND
28 16MM CLEAR PAIRED GLASS
29 LIGHT GAUGE STEEL FRAME CEILING RIB(TH-BAR) / T15 MITONE
30 GAUGE STEEL COLUMN 1: BANK, FIREPROOF, OIL PAINT(APPOINTED COLOR, FLATTING, 1 HOUR FIREPROOF)
31 C-100X50XT2.3
32 T10 AL. HONEYCOMB PANEL W/PANEL FIXATION HARDWARE(PRIVATE)
33 T70 INSULATION
34 T6 PL : BANK, OIL PAINT(APPOINTED COLOR, FLATTING) W/M16(F 10T)
35 SPEAKING ROOM
36 URETHANE FOAM FILLING
37 Ø -150X150XT5

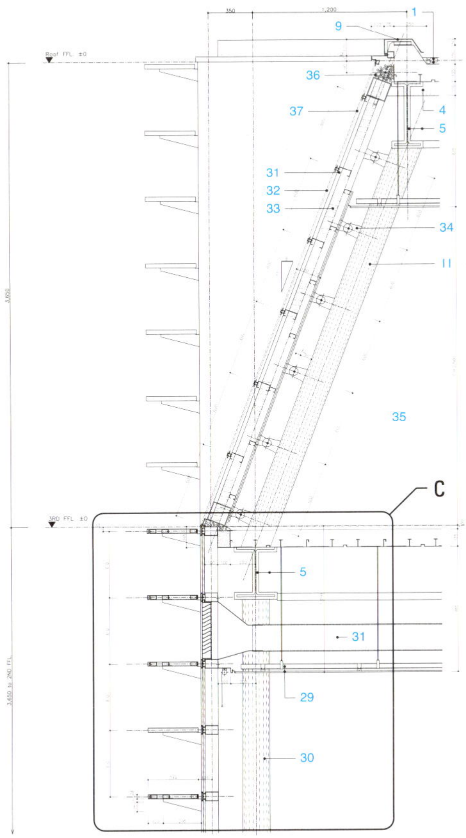

LOUVER SKIN WALL SECTION B

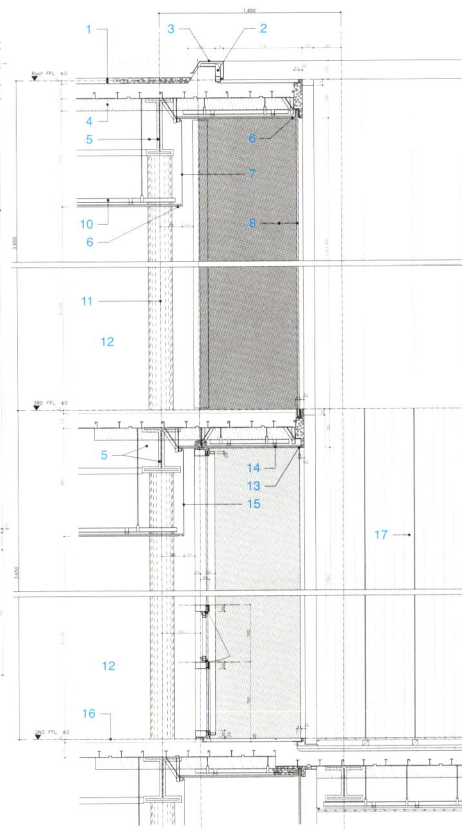

OUTER WALL SECTION D

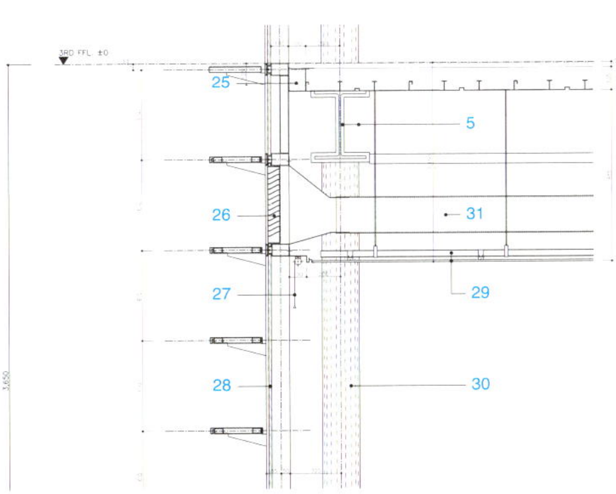

LOUVER SKIN DETAIL C

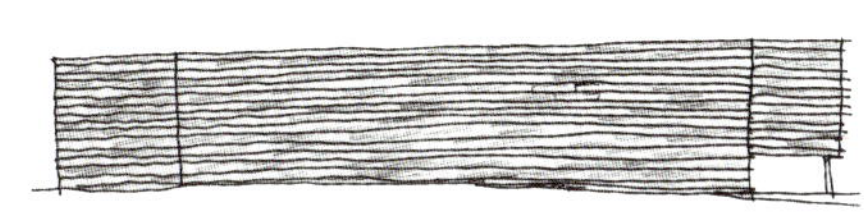

LOUVER SKIN SKETCH

COLOR GLASS COLUMN SKETCH

Multiple space composition through diverse level(program plaza)

Color glass column detail

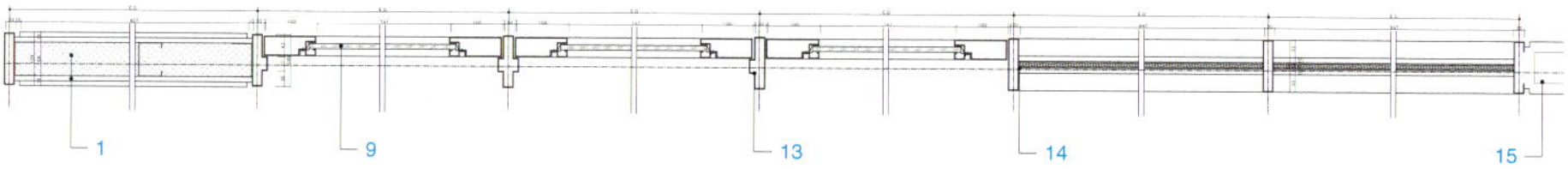

SECTION G

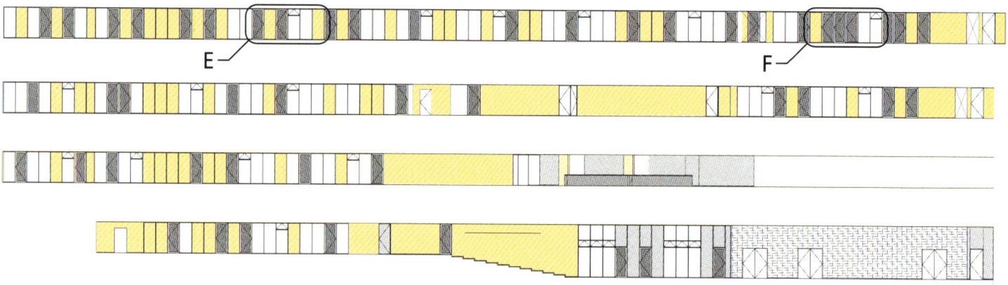

CORRIDOR DEVELOPMENT FIGURE

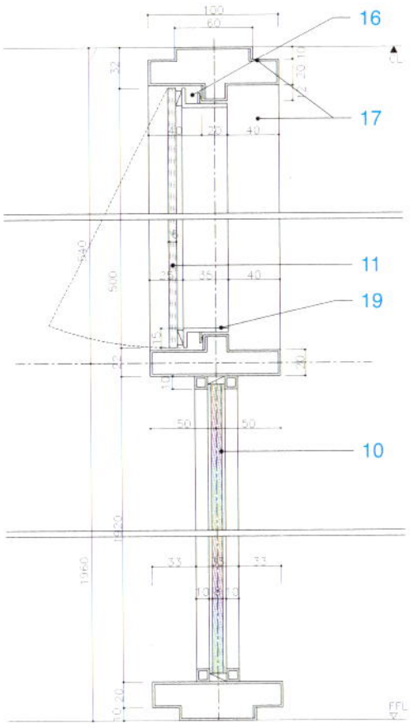

SECTION H

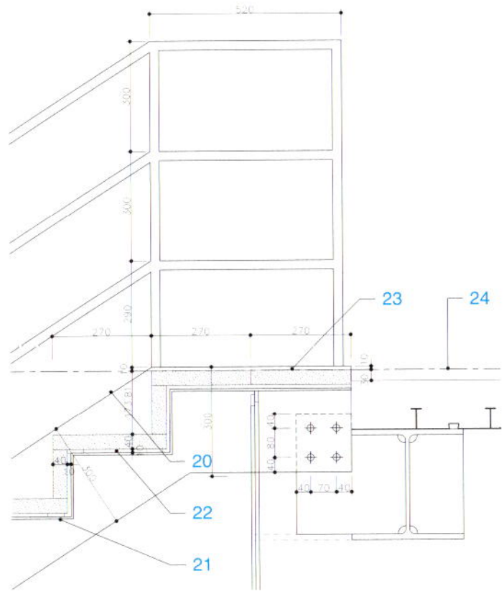

DETAIL I

1 C-STUD(65X45XT0.8, HOR. @450) W/T50 GLASS / VINYL PAINT ON T9.5 GYPSUM BOARD TWOFOLD AFTER ALL PUTTY
2 BASE BOARD : MIXED PAINT ON T9 PANEL
3 LIGHT GAUGE STEEL FRAME CEILING RIB / VINYL PAINT ON T9.5 GYPSUM BOARD AFTER ALL PUTTY TWOFOLD
4 T50 GLASS / T1.6 GALVALUME STEEL PLATE FOLDING W/OIL PAINT(FLATTING) TWICE
5 15X130 MDF
6 T30 GOHONG STONE FLOOR RUBBING
7 LEVER TYPE HANDLE
8 T0.8 GALVALUME STEEL PLATE FOLDING / OIL PAINT(FLATTING)TWICE
9 PC-PANEL(T10 ABS SHEET)
10 T8 TEMPERED GLASS W/ETCHING FILM
11 T6 TEMPERED GLASS
12 PL T10(H=100),BANK, OIL PIANT
13 T1.6 GALVALUME STEEL PLATE / OIL PAINT (FLATTING) TWICE
14 CAULKING(RIGHT, LEFR PER 4)
15 STRUCTURE TYPE, LIGHT GAUGE WALLS-STUD
16 PROJECT WINDOW HARDWARE
17 T1.6 GALVALUME STEEL PLATE FOLDING / OIL PAINT
18 STRINGER(PL T30, W=300),BANK PAINT, OIL PAINT
19 T3 STEEL PLATE CUTTING(OPEN ONLY FRAME)
20 STRINGER(PL T30 / W=300, BANK, OIL PAINT)
21 PL T6 BANK, OIL PAINT
22 T10 STICKED MAT FIXING
23 T40 WOOD FLATTING CLEAR URETHANE COATING
24 DECO TILE ON CEMENT MORTAR
25 Ø -80X40XT2.3 W/BANK(@450) / T12 WATER-PROOF PLYWOOD / T20 IPE WOOD

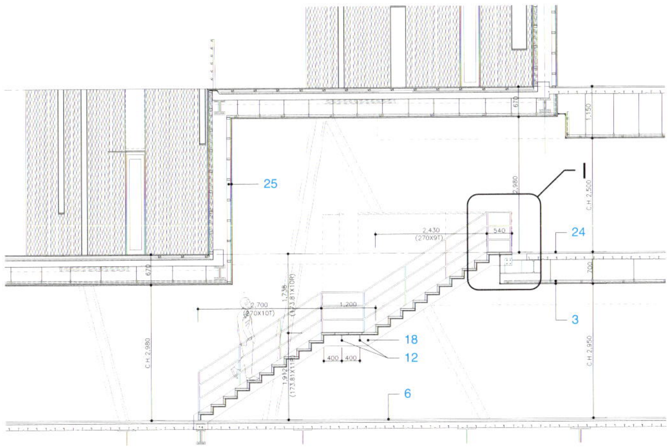

STAIR CROSS SECTION

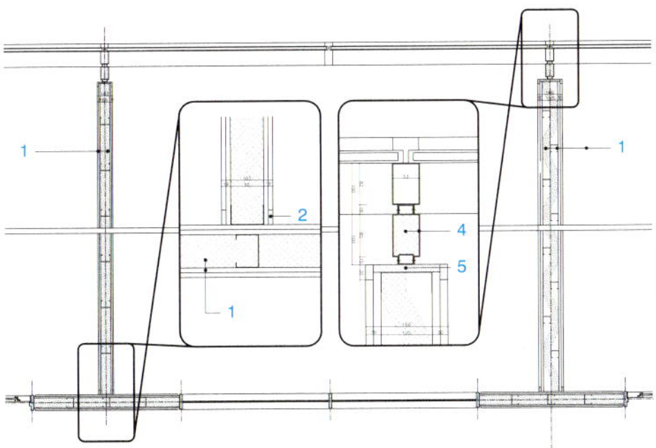

SECTION E

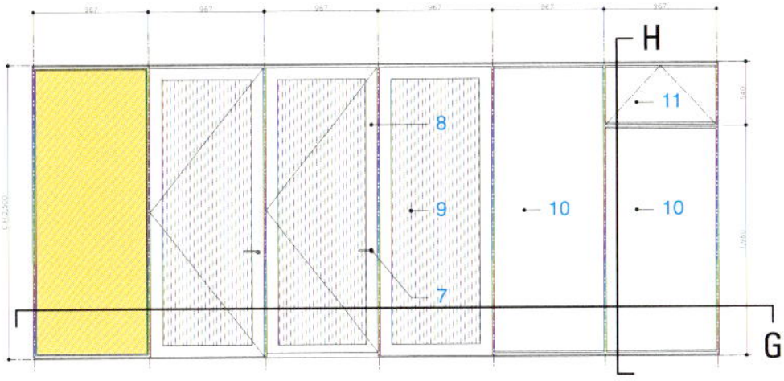

DETAIL F

The space encircling the building

1 CLASSROOM
2 TEACHER'S ROOM
3 LABORATORY
4 STUDY HALL
5 HALL
6 PARKING LOT
7 EQUIPMENT ROOM
8 CAFETERIA
9 LOUNGE
10 OFFICE
11 FUNCTIONAL ROOM

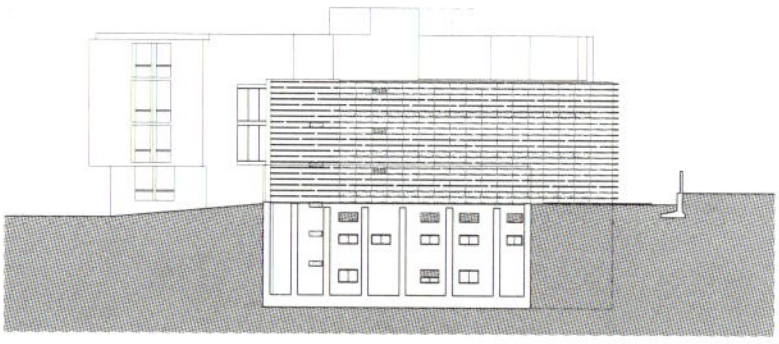

ELEVATION J

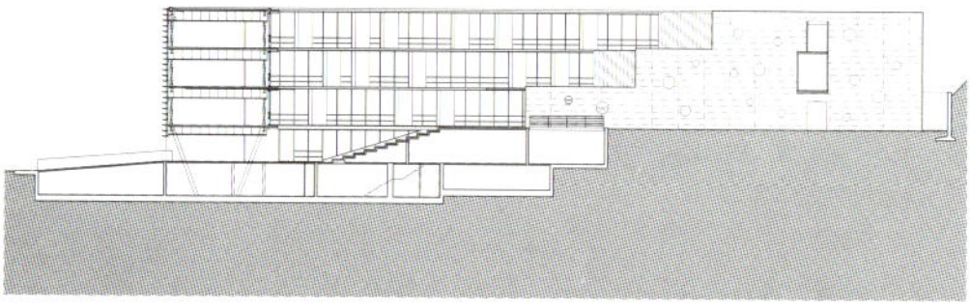

CROSS SECTION L

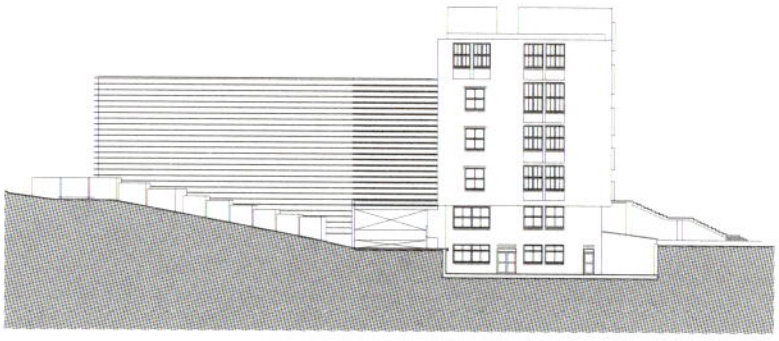

ELEVATION K

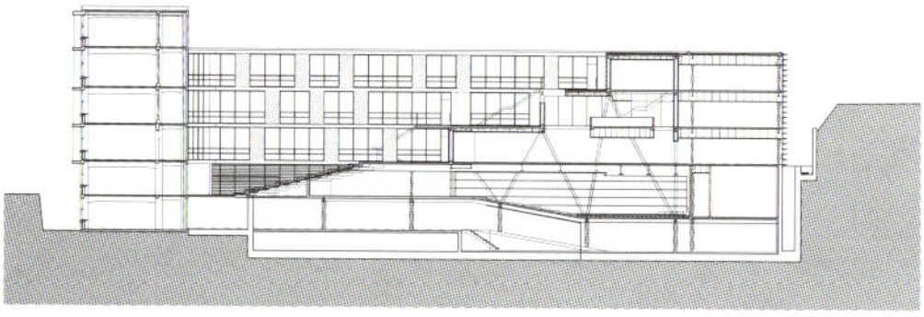

CROSS SECTION M

Upper part(lounge) of the cafeteria viewed from the laboratory on the 2nd floor

The cafeteria on the 1st floor

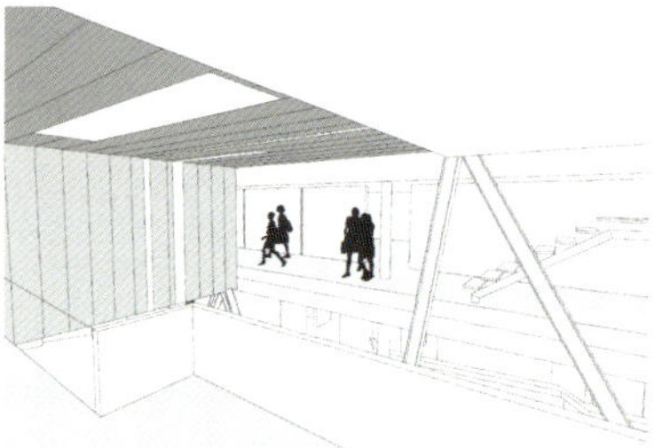

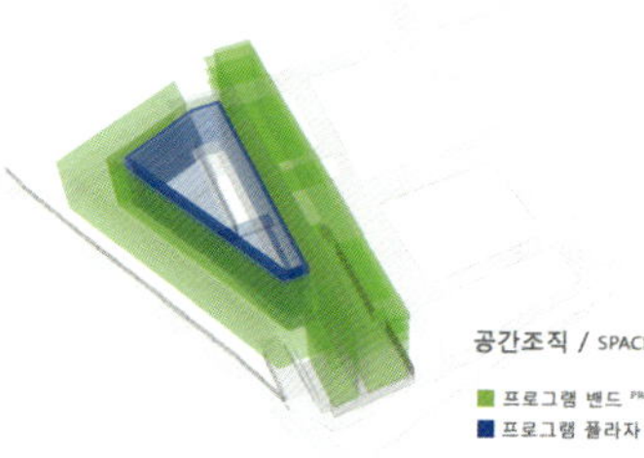

공간조직 / SPACE LOGIC

- 프로그램 밴드 PROGRAM BAND
- 프로그램 플라자 PROGRAM PLAZA

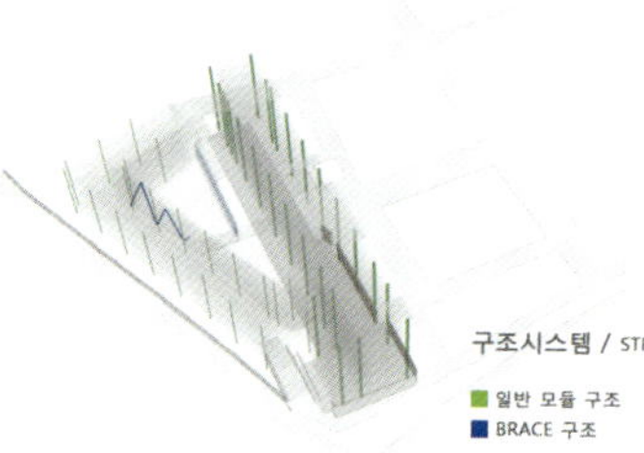

구조시스템 / STRUCTURE SYSTEM

- 일반 모듈 구조
- BRACE 구조

SPACE LOGIC & STRUCTURE SYSTEM

Structure detail

>> EDUCATION

SEOWON UNIVERSITY ART CENTER

Location Mochung-dong, Heungdeok-gu, Cheongju-si, Chungcheongbuk-do, Korea **Site Area** 156,304m² **Total Floor Area** 6,355.839m² **Building Coverage Ratio** 18.04% **Floor Area Ratio** 59.61% **Building Scope** B1~4F **Structure** RC **Exterior Finish** Exposed Concrete, Cement Panel, Steel Curtain Wall, AL Curtain Wall, Basalt **Interior Finish** Exposed Concrete, Cement Board, Gypsum Board, Sound-Absorbing Panel **Design Period** 2005.12~2006.5 **Construction Period** 2007.7~2008.10 **Architecture Design** Kim Tae-jip, Cho Joung-ho | GANSAM Partners Architects & Associates **Design Team** Kim Hyeon-ju, Seong Chi-ho, Kang Dong-gyun, Lee Dong-eun, Na Seung-mun **Interior & Landscape Design** GANSAM Partners Architects & Associates **Photographer** GANSAM Partners Architects & Associates

General view from the east

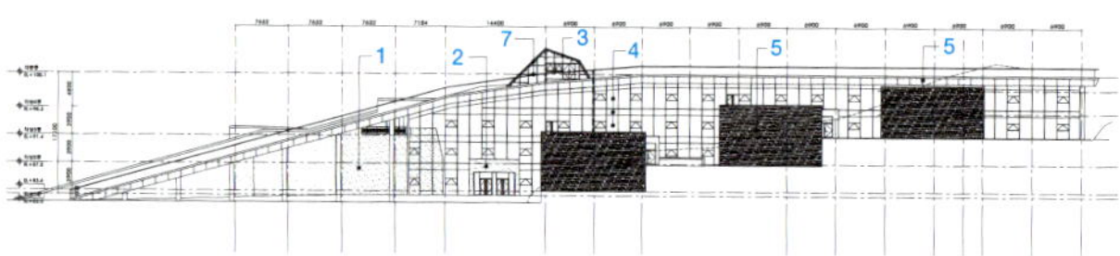

NORTH ELEVATION

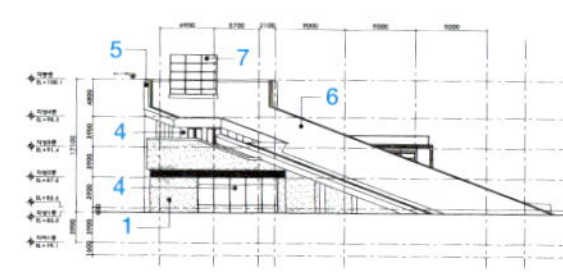

EAST ELEVATION

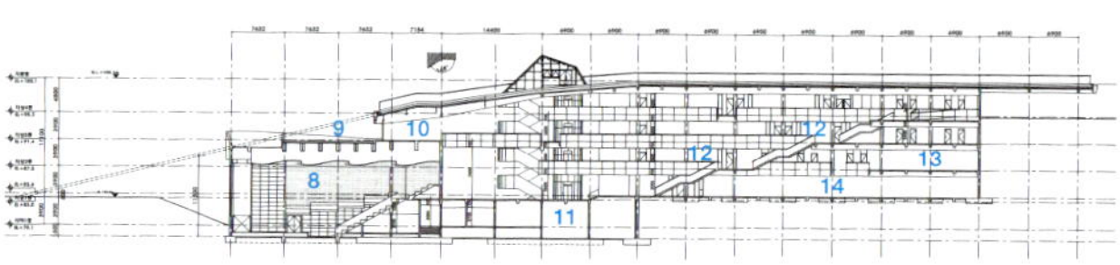

CROSS SECTION A

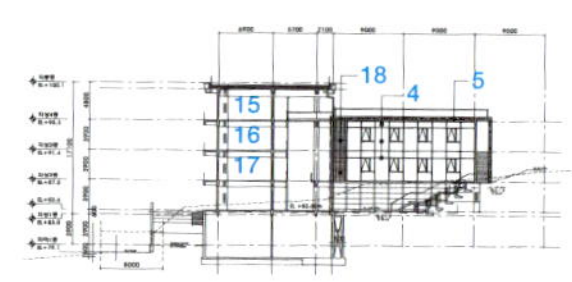

LONGITUDINAL SECTION B

The art center of Seowon University was designed to take a role of a bridge that visually and spatially connects natural greens, Mirae square that is a main space for pedestrians in the campus and the southern residential area. It brings the natural green that is in the furthest part from the front entrance into the campus under the concept of Eco Bridge. The big southern elevation that is opened toward the residential area(city) also functions as a Hyper Bridge to express a developed future image of Seowon University.

There is a performance hall that accommodates about 300 people on the left side of the lobby on the 1st floor and an atrium with stairs that connects four floors from the 1st to the 4th and a corridor on the right side. There are also a public zone and a practice zone facing each other over this atrium. The atrium takes a role of a spandex and three practice houses are connected along the level of natural greens around it. Natural greens are connected with the courtyard surrounded by these three practice houses to give natural ventilation and natural lighting functions to practice rooms and help students have a space for a rest and meeting in nature. The basement consists of a performance hall that is opened toward the south through a sunken space, a machinery room, a switchboard and others.

CAMPUS AND ART CENTER The concept of an eco - bridge that connects nature with the campus cores was introduced and the big southern elevation, which is opened toward the city, will function as a hyper bridge to express developed future images of Seowon University.

ATRIUM AND ELEVATION The topographic flow was reflected on the interior of the building and various spaces were formed with stairs. The vertical bar created by the column of pillars and a steel curtain wall emphasizes the spatial flow and transparent glass curtain walls were used as an elevation material for the visual connection between an art courtyard and a music courtyard.

PERFORMANCE HALL AND LOBBY To emphasize the image of stairs as an art piece in the lobby, stairs and the roof floor were designed to be attached to the girder on the 3rd floor by a tie bar.

MUSIC COURTYARD The topography with different levels forms dynamic atmosphere and the hall that surrounds the rhythmical garden helps visitors aggressively experience the courtyard. Text by GANSAM Partners Architects & Associates

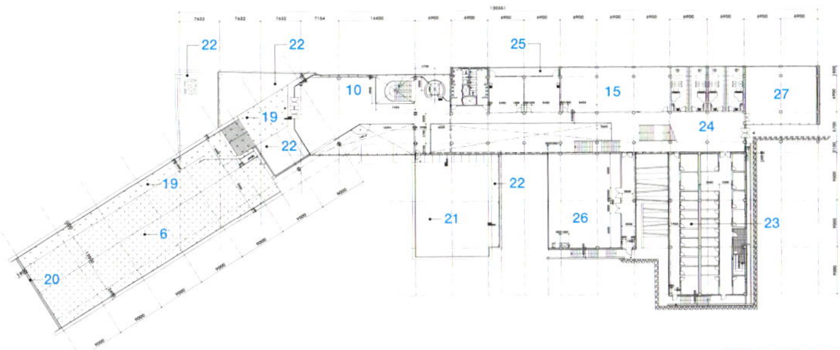

3RD FLOOR PLAN

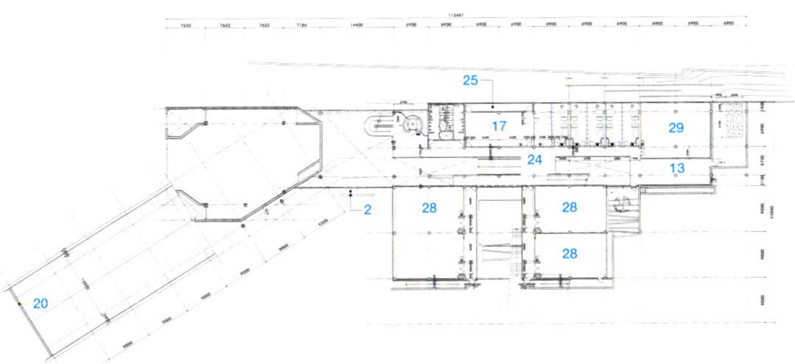

2ND FLOOR PLAN

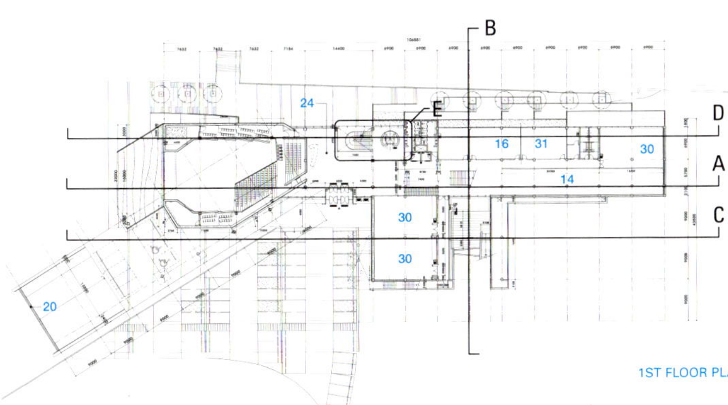

1ST FLOOR PLAN

1 T30 BASALT
2 EXPOSED CONCRETE
3 T4 ALUMINUM COMPOSITE PANEL
4 T22 CLEAR PAIRED GLASS
5 EXPOSED CONCRETE / TRANSPARENT WATER REPELLENT
6 SEDUM ON T100 ARTIFICIAL SOIL (LANDS- CAPING WORK)
7 T28.76 CLEAR LAMINATED SEMI-TEMPERED GLASS
8 PERFORMANCE AREA
9 DECK
10 MULTIPURPOSE HALL
11 MACHINERY ROOM
12 ATRIUM
13 REST AREA
14 EXHIBITION ROOM
15 KOREAN MUSIC ROOM
16 LECTURE ROOM
17 CLASSROOM FOR PRINT
18 T50 EXTRUDED FORMED CEMENT PANEL
19 T50 GRASS BLOCK (LANDSCAPING WORK)
20 W500 GALVANIZED GRATING COVER TRENCH
21 T60 SPECIFIC CONCRETE BLOCK
22 CLEAR HARDENER ON PLAIN CONCRETE FIN
23 CLASSROOM FOR LESSON
24 HALL
25 BALCONY
26 CONCERT ROOM
27 PERFORMING ROOM
28 CLASSROOM FOR WESTERN PAINTING
29 CLASSROOM FOR ROUGH DRAWING
30 CLASSROOM FOR ORIENTAL PAINTING
31 CLASSROOM FOR VIDEO-PAINTING

Southern elevation

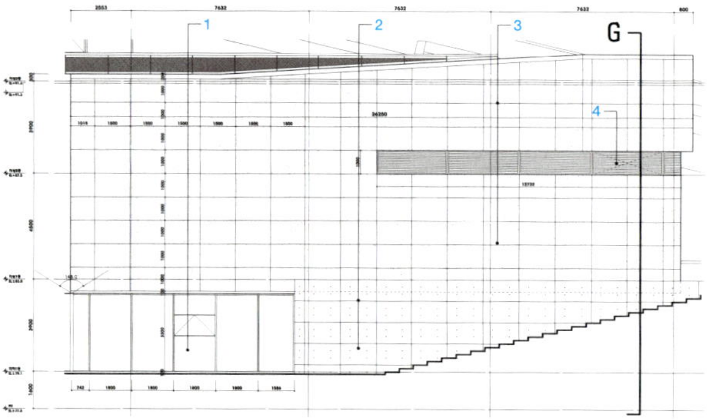

OUTER WALL ELEVATION F

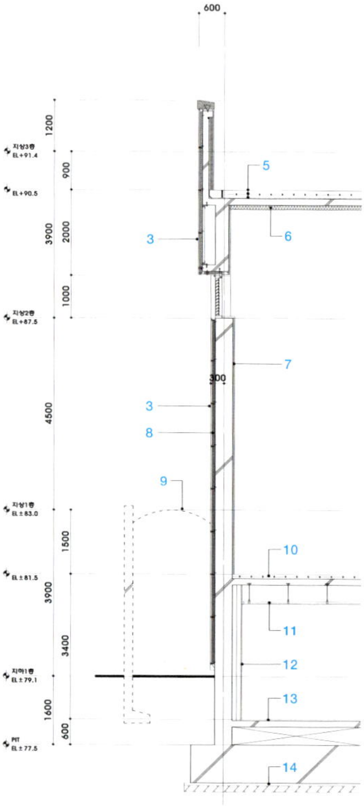

OUTER WALL SECTION G

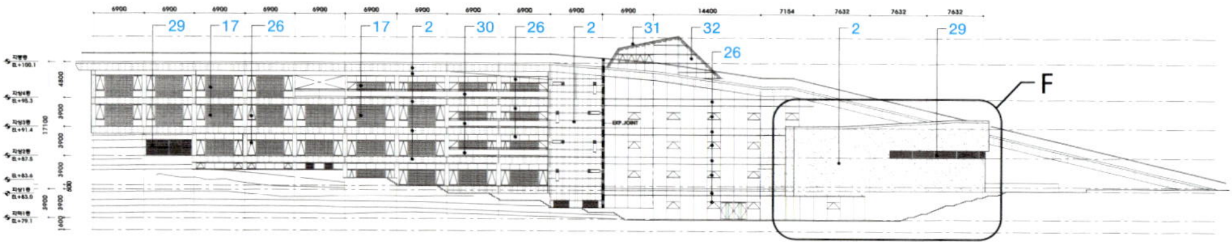

SOUTH ELEVATION

1 T22 COLOR PAIRED GLASS
2 EXPOSED CONCRETE / TRANSPARENT WA-TER REPELLENT
3 T30 BASALT FIN
4 ALUMINUM GRILL
5 T120 PLAIN CONCRETE CLEAR HARDENER / WIRE MESH / T50 SLOPED MORTAR / SURFACE COATING WATERPROOFING, PROTECTIVE MORTAR
6 T110 COMPRESSED STYROFOAM INSULA-TION
7 T25 SOUND-ABSORBING PANEL
8 T65 COMPRESSED STYROFOAM INSULA-TION
9 FLOWER BED(LANDSCAPING WORK)
10 T0.3 EPOXY PAINT / PLAIN CONCRETE
11 T12 ROCK WOOL SOUND-ABSORBING TEX
12 WATER-BASED PAINT
13 T3 VINYL TILE
14 T60 LEVELLING CONCRETE / T150 CODDLE STONE COMPACTION
15 T100 ARTIFICIAL SOIL, SEDUM(LANDSCAPING WORK) / T100 DRAIN LEVEL(LANDSCAPING WORK) / T60-120 SLOPE CONCRETE / SURFACE COATING WATERPROOFING, PROTECTIVE MORTAR
16 T9 CEMENT BOARD, CLEAR WATER REPELL-ENT
17 T50 EXTRUDED FORMED CEMENT PANEL
18 EXPOSED CONCRETE SURFACE TREATMENT
19 T9 POLISHING TILE
20 GYPSUM BOARD, WATER-BASED PAINT
21 W300 U-TYPE GUTTER
22 T50 BEAD METHOD INSULATING BOARD (NO.1) / T60 LEVELING CONCRETE / T150 CODDLE STONE COMPACTION
23 EXPANSION JOINT(2700X2700)
24 SURFACE COATING WATERPROOFING, PRO-TECTIVE MORTAR / CEMENT BLOCK UPRIGHT LAYING / CEMENT MORTAR
25 T3.0 ALUMINUM SHEET / FLUOROCARBON RESIN COATING
26 T22 CLEAR PAIRED GLASS
27 STEEL CURTAIN WALL
28 CURTAIN BOX T:1.2 STL PL / ELECTROSTATIC POWDER COATING
29 ALUMINUM LOUVER
30 T10 OPAQUE TEMPERED GLASS
31 T4 ALUMINUM COMPOSITE PANEL
32 T28.76 TRANSPARENT LAMINATED SEMI-TEMPERED GLASS
33 CLASSROOM FOR WESTERN PAINTING
34 CONCERT ROOM
35 COURTYARD
36 LESSON ROOM
37 CLASSROOM FOR ORIENTAL PAINTING

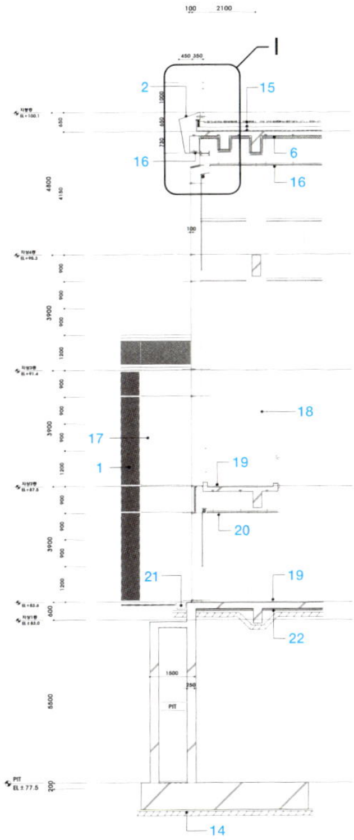

OUTER WALL SECTION H

Void staircase

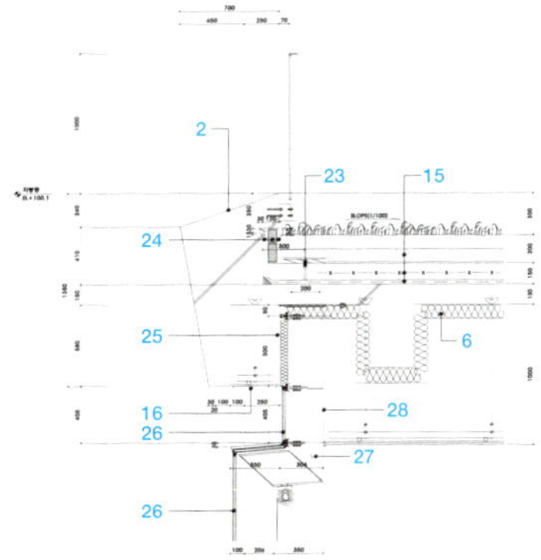

ROOF DETAIL I

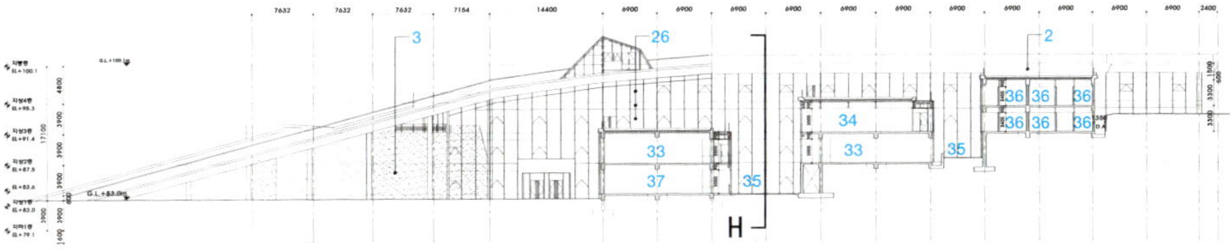

CROSS SECTION C

Core penetrated from B1 floor until 4th floor

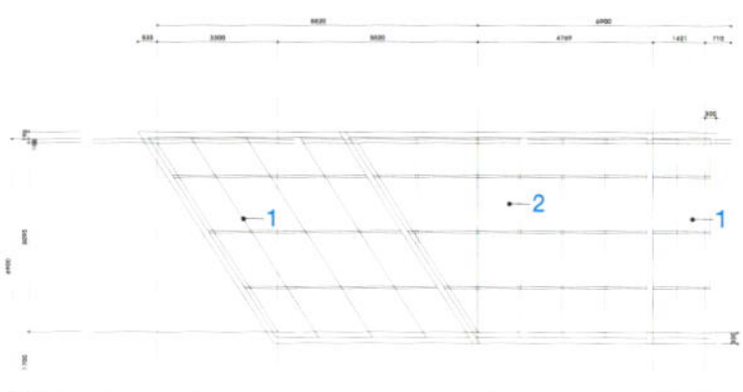

CORE SKYLIGHT FLOOR PLAN

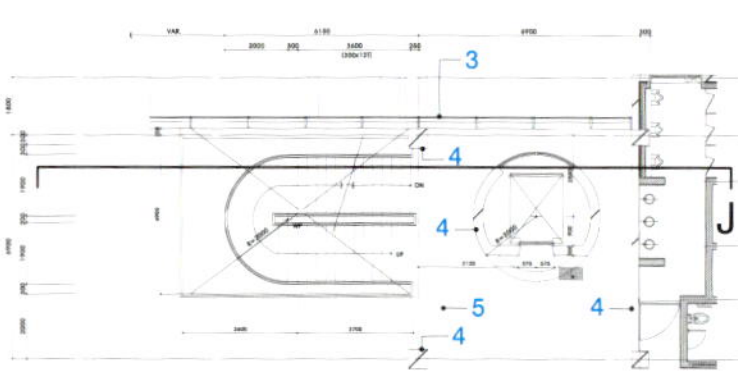

CORE FLOOR PLAN E

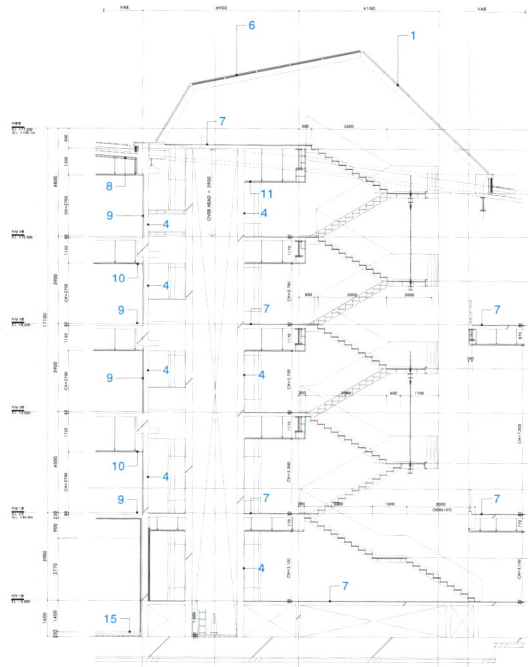

SECTION J

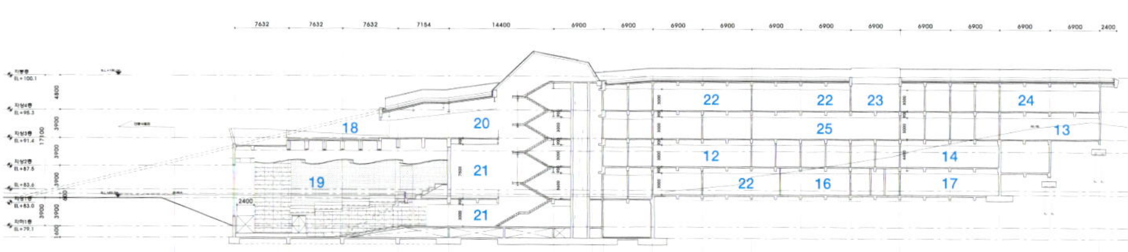

CROSS SECTION D

1 T28.76 CLEAR LAMINATED SEMI-TEMPERED GLASS
2 T4 ALUMINUM COMPOSITE PANEL
3 T22 CLEAR PAIRED GLASS
4 EXPOSED CONCRETE
5 T9 POLISHING TILE
6 T4.0 ALUMINUM COMPOSITE PANEL
7 CEMENT MORTAR / T9 POLISHING TILE
8 T110 COMPRESSED STYROFOAM
9 CEMENT MORTAR / T7 PORCELANEOUS TILE
10 GYPSUM BOARD / VINYL PAINT
11 T9 CEMENT BOARD / CLEAR WATER REPELLENT
12 CLASSROOM FOR PRINT
13 PERFORMANCE ROOM
14 CLASSROOM FOR ROUGH DRAWING
15 T0.3 EPOXY PAINT / T200 PLAIN CONCRETE / INFILTRATION WATERPROOFING
16 CLASSROOM FOR VIDEO-PAINTING
17 CLASSROOM FOR ORIENTAL PAINTING
18 DECK
19 PERFORMANCE AREA
20 MULTIPURPOSE HALL
21 HALL
22 LECTURE ROOM
23 REST GARDEN
24 ROOM FOR LISTENING TO MUSIC
25 KOREAN MUSIC ROOM

View from the staircase

The performance area on the B1 floor

The performance area interior viewed from the stage

원각도서관

GEUMGANG UNIVERSITY CENTRAL LIBRARY

Location Daemyeong-ri, Sangwol-myeon, Nonsan-si, Chungcheongnam-do, Korea Site Area 123,894.00m^2 Building Area 2,504.00m^2 Total Floor Area 12,067.80m^2 Building Coverage Ratio 2.02% Floor Area Ratio 5.74% Building Scope B2~4F Maximum Height 24.3m Structure RC Exterior Finish T12 Clear Laminated Glass(Louver), T30 Granite Smooth Trimming, T18 Clear Pair Glass Design Period 2004.4~2005.1 Construction Period 2005.2~2006.12 Architecture Design Im Sung-pil | JUNGLIM Architecture Design Team Son Byeong-chan, Yun Cang-min, Hwang Jae-yeon, Yun Seong-ho Photographer Park Wan-soon

PLACING IT AT THE PROPITOUS SITE It is a propitous site located at the edges of Mt. Kyeryong, facing Mt. Noseong in Daemyeong-ri, Sangwol-myeon, Nonsan-si, Chungcheongnam-do, where Sangwoljosa who established Cheontaejong positioned himself there to get nirvana. It is surrounded by agricultural fields and farms, and a small temple and a lake on the left side reflect the clam but energetic nature.

SYMBOL The building of Wongak academy and information hall, GeumGang University symbolizes Buddhism which represents the school and the sect. Its straight symbolism is reflected in the lotus-shaped roof, which is decisive in the configuration of the whole campus and the prefect round form represents the optimal state that Buddhism ultimately pursues. The overall horizontal segmentation and vertical louvers visually create exceptional scenery and neutral atmosphere with the natural light. Such Buddhistic symbolism is harmonized with the surroundings to create the landscape and the nature does not customize the building but embrace it as a part of the landscape.

CAMPUS CORE Wongak academy and information hall of GeumGang University was designed on the premise of GeumGang University master plan. The exterior is visually and spatially consistent with the adaptable topology, access to the multi levels through the slant, and the lecture rooms which are adjacent by podium plan.

VARIABLE FACADE The glass louver on the elevation of the building represents frontage of the three sides and transparency of the building. Especially, the angles of the louver vary the natural lights to be provided to maintain bookstands in the library while providing indirect light for reading. In addition, the louver creates ever - changing elevation images for different seasons.

ROOF ORNAMENT / PENETRATING SPACE The roof structure, which materializes a lotus at the center of the building, penetrates the inner court and creates a meditative space in the Wongak Hall, the center of the building. As a place accepting the natural light, it accommodates students' movements over the levels and different activities. Text by JUNGLIM Architecture

Front view

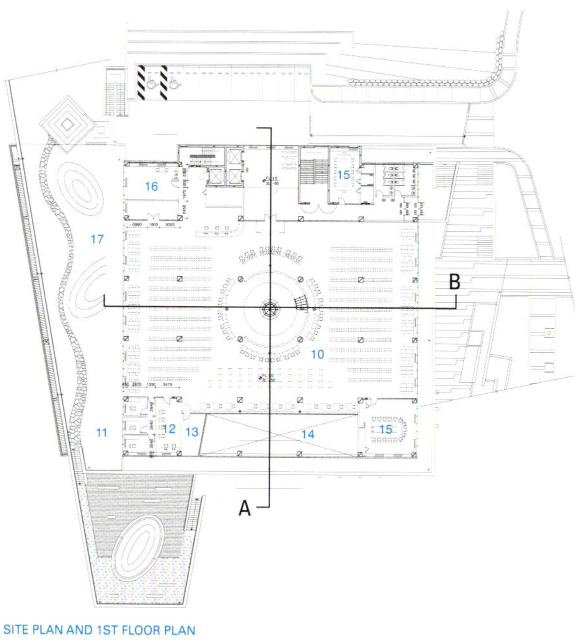

SITE PLAN AND 1ST FLOOR PLAN

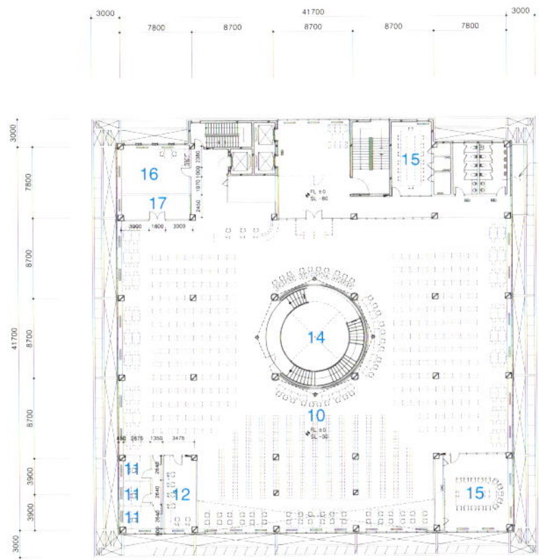

2ND FLOOR PLAN

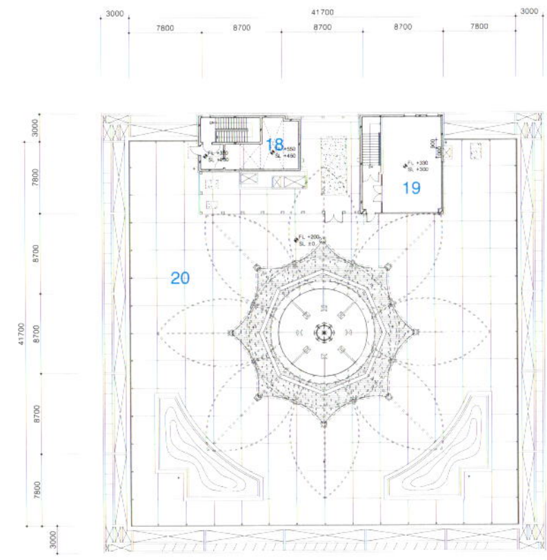

ROOF PLAN

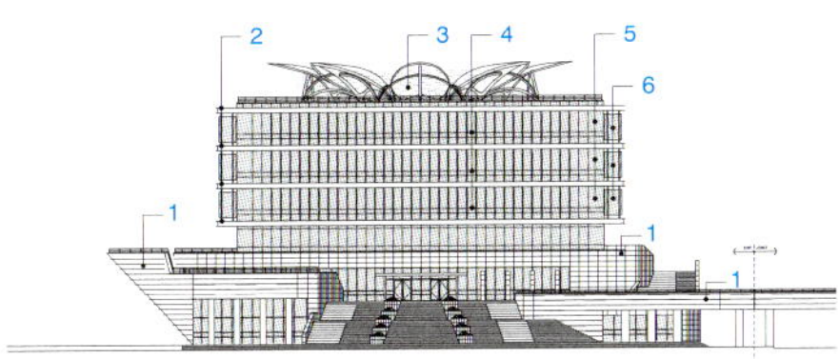

FRONT ELEVATION

REAR ELEVATION

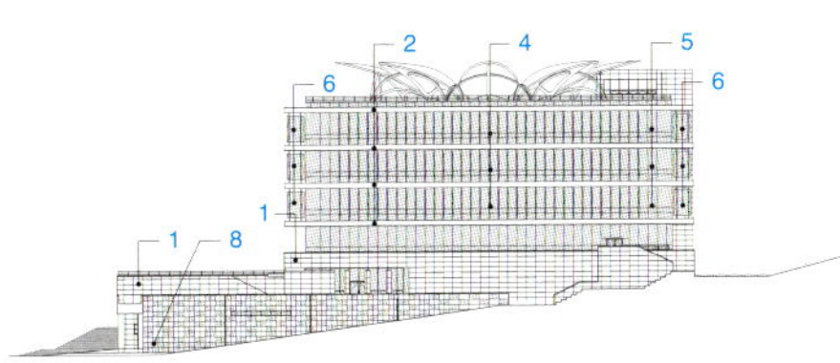

RIGHT ELEVATION

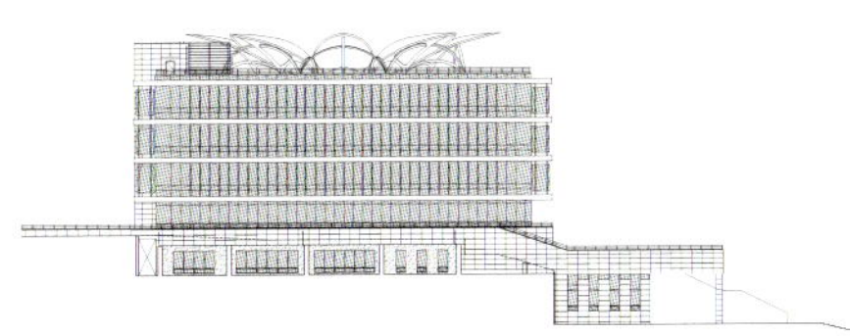

LEFT ELEVATION

1 T30 GRANITE-A
2 ⊏-600X200X12 STL / URETHANE PAINT
3 ALUMINUM SQUARED PIPE 100X50(@300)
4 T12.76 TEMPERED, LAMINATED GLASS-(LOUVER)
5 T18 COLOR PAIRED GLASS
6 T12.76 TEMPERED, LAMINATED GLASS
7 T100 OUTSIDE INSULATING SYSTEM
8 T30 MARBLE STONE SMOOTH TRIMMING
9 T18 CLEAR PAIR GLASS
10 OPEN READING ROOM
11 CAPSULE ROOM
12 SMALL GROUP WORKROOM
13 REST ROOM
14 OPEN
15 SEMINAR ROOM
16 ACQUISITION WORKROOM
17 COPYROOM
18 ELEVATOR MACHINERY ROOM
19 WATER TANK ROOM
20 ROOF GARDEN

The upper wall viewed from the lower zone

1 OPEN READING ROOM
2 LIBRARIAN ROOM
3 DISABLED PERSON READING ROOM
4 LOBBY
5 AUDIOVISUAL ROOM
6 REST ROOM
7 HALL
8 T3.0 ALUMINUM SHEET
9 T60 INSULATION
10 Ø100 STL PIPE / ELECTROSTATIC POWDER COATING
11 T12.76 TEMPERED LAMINATED GLASS-(LOUVER)
12 TGP SYSTEM
13 FCU COVER
14 URETHANE PAINT ON T1.2 STP / GALVANIZED □ - 40X40X2.3
15 H-588X300X12X20 / URETHANE PAINT
16 SPREADING Ø25 BLACK-BEAN GRAVEL
17 T30 GRANITE RUBBING
18 T65 INSULATION / T0.03 PE FILM TWOFOLD / T60 LEVELING CONCRETE
19 PARAPET RING
20 0.5B PROTECTIVE BRICK / MORTAR / WATER-BASED PAINT
21 STL- Ø10 CIRCULAR BAR / URETHANE PAINT @200
22 25X25 CAULKING / EXPANSION JOINT @3000X3000 / W=25 JOINT FILLER
23 ⊏- 600X200X12 STL / URETHANE PAINT
24 T18 CLEAR PAIRED GLASS
25 T1.2 RUST-PREVENTING STEEL PLATE, MIXED PAINT / T50 GLASS WOOL
26 □-50X30X2.9 / T12 WATERPROOF PLYWOOD
27 SET ANCHOR
28 BACK-UP MATERIAL, CAULKING
29 50X50X5T STEEL WASHER
30 M12X80LG ANCHOR BOLT SET
31 CEMENT MORTAR / WATER-BASED PAINT FOR OUTSIDE
32 10X10 MATERIAL SEPARATOR GROOVE DIGGING
33 FILLING MINERAL WOOL
34 M12 ANCHOR BOLT SET
35 2L-100X110X6T STEEL BRACKET
36 T1.2 ST'L
37 M12X100LG BOLT, NUT / SPRING WASHER
38 40X40X5T STEEL WASHER
39 T1 VINYL SHEET

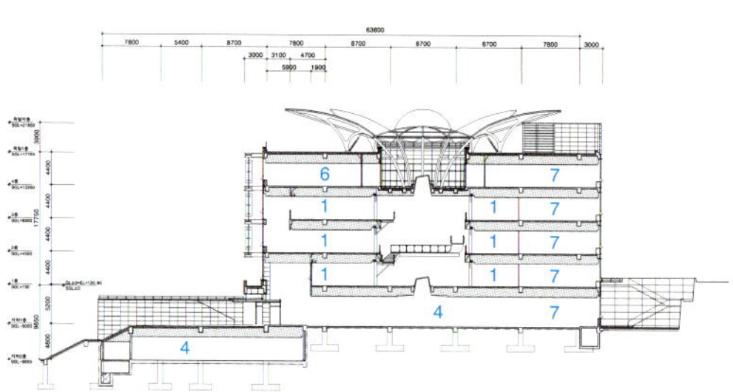

LONGITUDINAL SECTION A

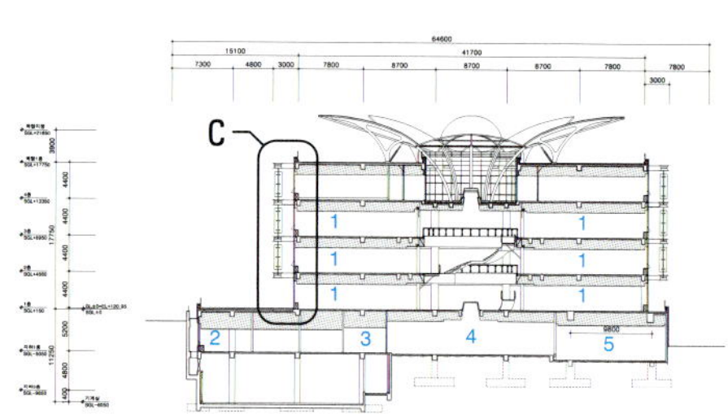

CROSS SECTION B

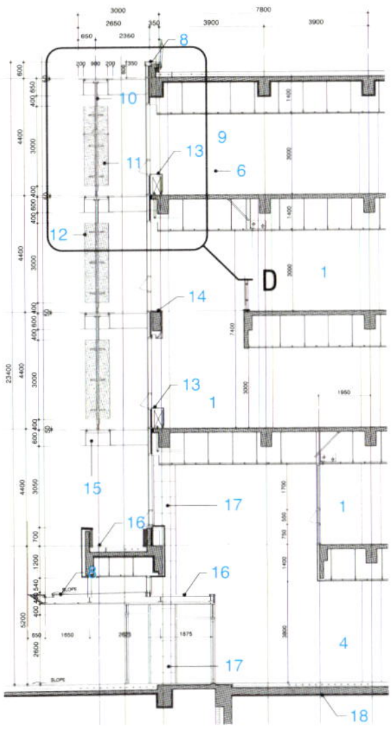
OUTER WALL DETAIL C

Glass louver detail

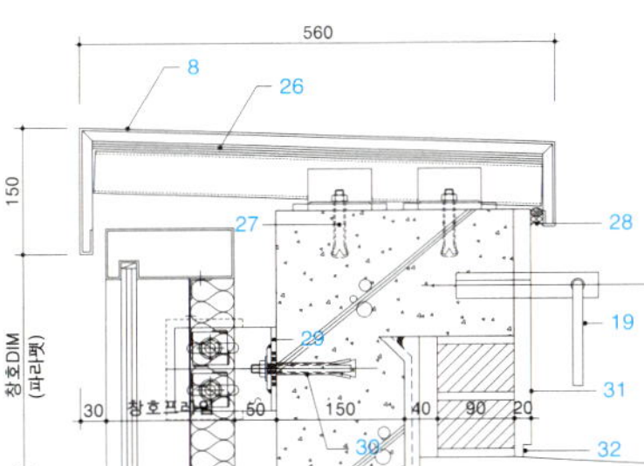

DETAIL E

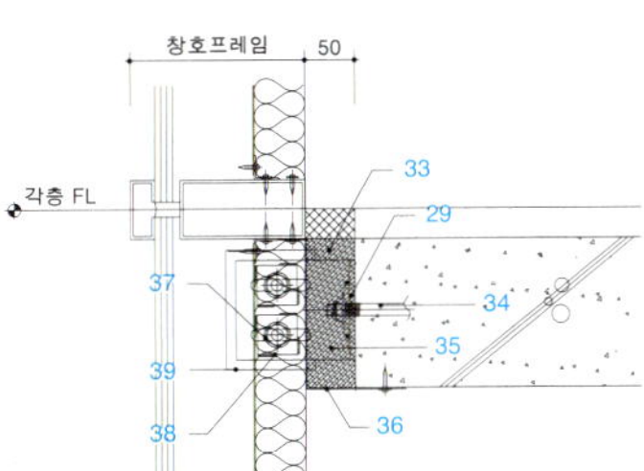

DETAIL F

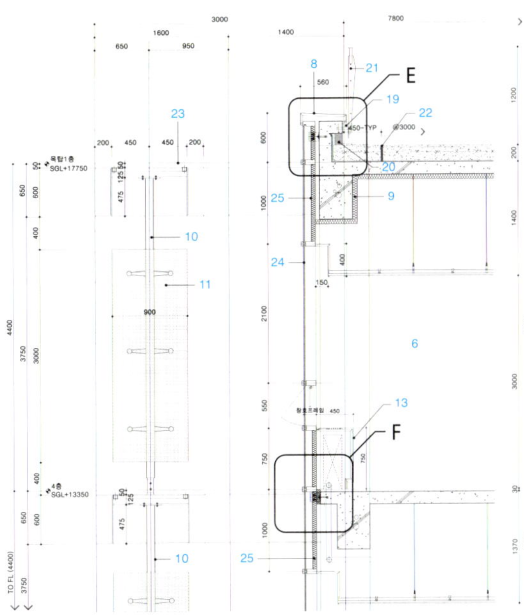

GLASS LOUVER DETAIL D

Lotus-shaped roof structure

The meditation space of the wongak hall

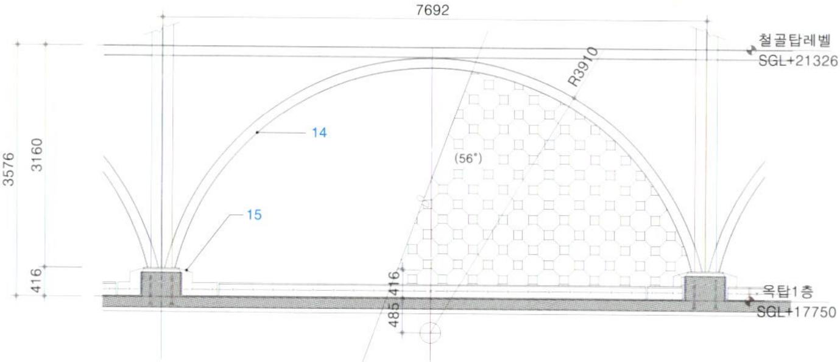

ROOF FRAME SECTION G

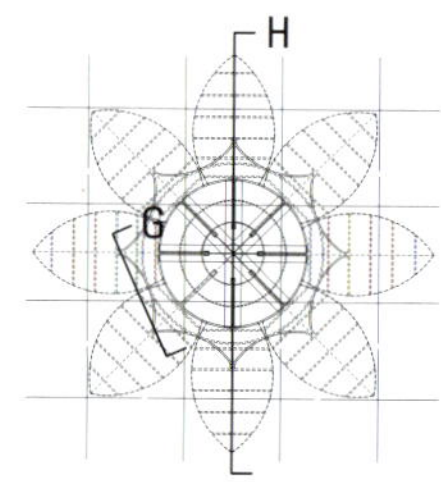

KEY PLAN

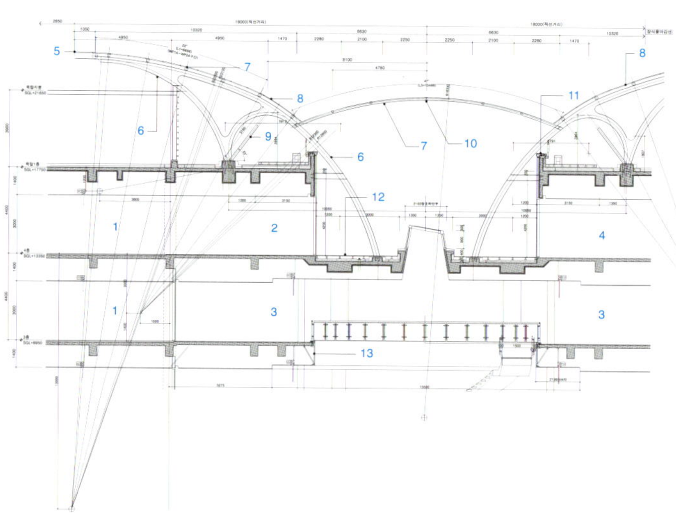

ROOF FRAME SECTION H

1 ELEVATOR HALL
2 LOBBY
3 OPEN READING ROOM
4 REST AREA
5 PL-200X4.5T / URETHANE PAINT
6 STEEL ELEMENTS / URETHANE PAINT
7 □-150X150X6 @2000
8 T3 PERFORATED AL.SHEET / ELECTROSTATIC POWDER COATING / □-40X40X2@900X1200
9 STEEL SLOPED ELEMENTS
10 H-150X150X7X10
11 T3 ALUMINUM FLASHING
12 WOOD DECK
13 T1.2 STEEL PLATE / ELECTROSTATIC POWDER COATING
14 H-150X150X7X10 SLOPED ELEMENTS / URETHA NE PAINT
15 PLAIN CONCRETE WATERPROOF HUMP / ROOF DECORATION TYP

The circular staircase is located in the center of the interior.

Circular staircase

EWHA WOMANS UNIVERSITY EDUCATION BUILDING

Location Daesin-dong, Seodaemun-gu, Seoul, Korea **Site Attribute** City, General Resident, History-Cultural Aesthete **Site Area** 538,721.80m² **Building Area** 78,608.99m² **Total Floor Area** 369,000.29m² **Building Coverage Ratio** 14.59% **Floor Area Ratio** 53.98% **Building Scope** B2~6F **Structure** SRC **Exterior Finish** T30 Granite Smooth Trimming, T24 Clear Pair Glass, T3 Aluminum Sheet, Aluminum Adiabatic Bar **Interior Finish** Lobby_T30 Granite Smooth Trimming, T30 Granite Rubbing, T30 Western Red Cedar / Lecture Room_ Water-Based Paint **Architecture Design** Mon Jin-ho | JUNGLIM Architecture + Kim Ki-joong | 2105 Architects & Designers **Design Team** Yun Baek-ho, An Gwang-won, Nam Ji-hyeon, Kang Hyeong-ju **Construction** HYUNDAI Engineering & Construction **Photographer** Lee Ki-hwan

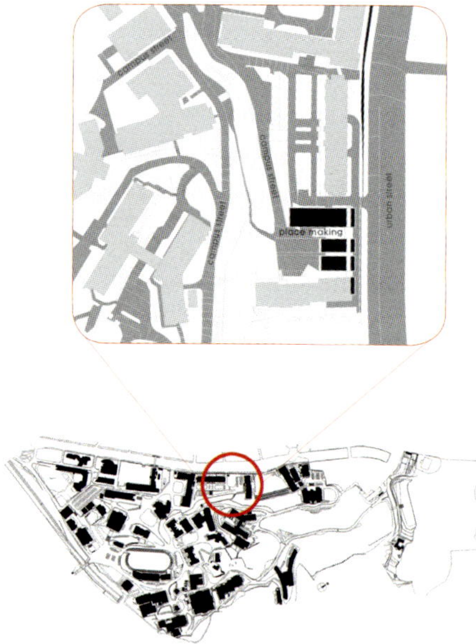

SITE PLAN

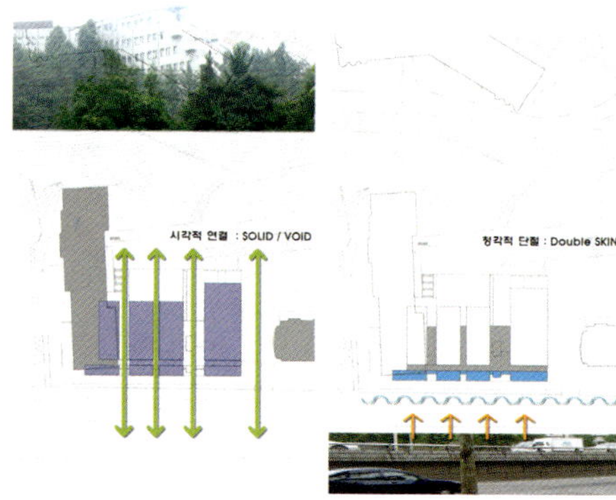

SITE RELATIONSHIP

MUTUAL ENTANGLEMENT OF PHENOMENON AND NOTION OF THE SITE

"The architecture is superior to geometry. It is a close combination between a notion and a shape. The meaning of the architecture exists in the mutual entanglement of the three, a phenomenon and a notion of the site which architecture belongs to." - S. Holl, 1983

ISOLATION & MUTUAL COMMUNICATION

- Place & Site: Ewha Woman's University has a geological characteristic which a slope is relatively steep and a rapid increase of the number of students brings a dense arrangement of the buildings in the campus. The site which the Education College holds its place is near the roadside in the rear gate of the campus and is located between the original Education College and the Humanities Building. It is a diagonal line, has a difference of level of about two stories and is located in the place to have the loudest street noise. Also, because of facing the north-west, its sunshine is poor. Meanwhile, the buildings of the Ehwa Womans University in the roadside stands in a long line like a wall along the street. It is an obstacle to block artificially the inside and the outside of the campus and a physical factor to violate a characteristic of the open university.
- Program: The new Education College is required for various programs owing to its characteristic and the facilities for every experiments and various events as well as the facilities for the lecture and the research are required. We divide these facilities into three categories, separate them three-dimensionally and make a zoning plan. We divide them into the lecture & research zone, the experiment zone and the multi zone and reflect it into a space and a shape plan.
- Design Process: First, the Education College should accept a program of about 10,000m². Seeing a size of an architecture, it is known that it is impossible to accept a side corridor and a duplicate corridor. Above all, beginning from the rectangular mass, we try to remove the outer space and divide a private use part and a common use part. Second, the Education College should contribute to mutual communication of the inside and outside of the campus between the original Education College and the Humanities Buildings. We think three different masses for a method for the mutual communication. The masses are connected mutually as a common space and then isolated soon. Such a connected common space is treated with glasses and utilized as a pathway of a mutual communication. Third, the Education College should execute a role to block a noise of the street and a sunshine. We think an idea to arrange a core space in the roadside for the method for it. Until an improvement of the wall of the original Education College, four core spaces are arranged as a corridor, an elevator and a toilet and execute their role for block. Fourth, the Education College should be in harmony with a connectivity and a hierarchy of the inner space. Seeing a campus constitution near the Education College, the original Education College, the Humanities Building and the Hakgwan are arranged like a 'ㄷ' and form a long outer space. The original outer space is used passively as a method for a passage rather makes a specific behavior spatially. Therefore, we plan to surround the outer space between the original Education College and the new one and it can be used in every events of it. In addition, the mass of the Education College is planned as a Piloty and considered to be possible to connect with the original outer space.

MODEL STUDY

MODEL-1

MODEL-3-1

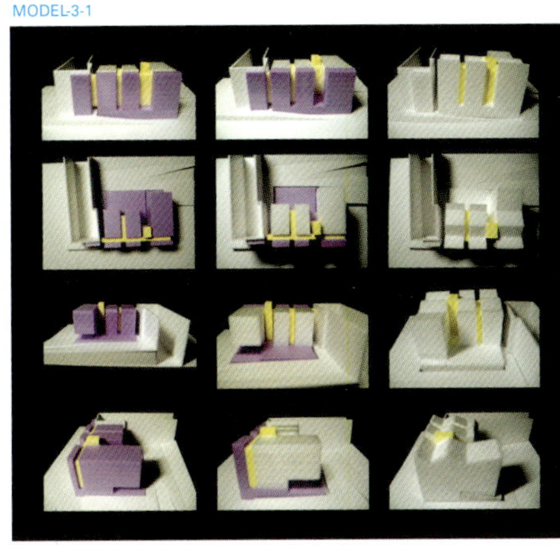

MODEL-2

MODEL-3

MODEL-4

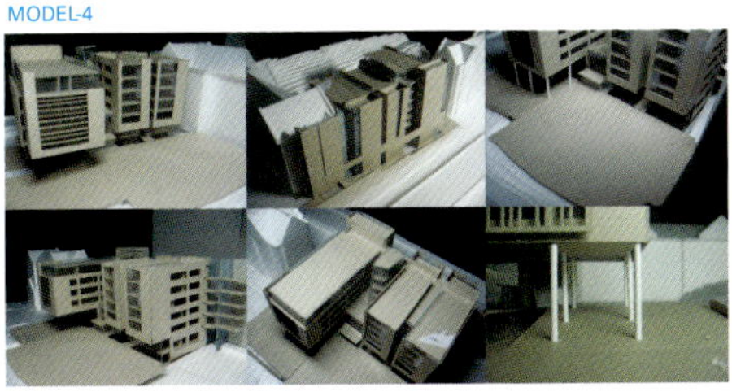

DIAGRAM & PROGRAM

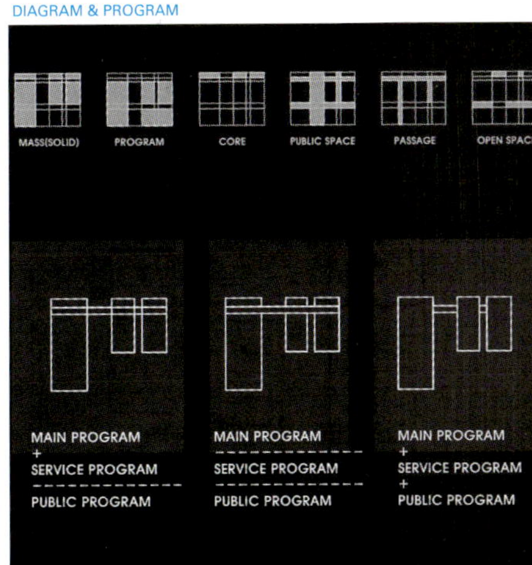

MODEL-FINAL

The elevation viewed from the road

MODEL STUDY

Model-1 The early model study to use Stryfoam is a work not only to verify a scale including an area but also to explain a relationship of the zoning of the program and the site.

Model-2 After going through the early mass study, we convert into a model study to use papers. We can go on a study about a structural problem, an area in the thought of mass, a progress of linear thought and a feeling of space for the outer design.

Model-3 In the third stage, the model to reflect an elevation design is made. At the time of early idea, an elevation of the building to face a big street is set as a full glass. It is planned to express a strong openness and at the same time to acquire a strong shield from a noise. But this idea faces a crisis due to an economy and a sunlight.

Model-3-1 We need a feedback for the progress of the third stage and decide to use a small mass model work in order to solidify a logic for the mass, type, program and elevation. These models enables us to experiment various ideas within a rapid time with a 1/600 scale. Particularly, we can arrange a concept of 'body, exterior and interior' for the elevation and the logic of the use.

Model-4 Through a previous work, in the fourth model study we tried to recover a handicap through zoning of program and it is a reasonable method in the economical and architectural side. Through a change of zoning of the program, we arrange the functions which correspond to the core in the roadside. By setting in an elevator, two corridors and a toilet, we protect lecture's rooms from sunlight and noise. For this, except for minimal windows for ventilation, we constitute a side of road with a concrete wall.

Model-Final In the final model, in order to reflect an elevation as well as its details, a model of 1/100 scale is born. In this scale, we can experiment a room for a windbreak room in the main entrance and an idea on the canopy. In particular, in this building, a design of the roof is also important, as the roof of the building is showed well in the campus. Hereby, the roof parapet is planned for a 600mm width wider than the general buildings. The matter to apply a logic of the elevation in fact is also important and in the parts to use stone, two modules and details are used and the same is at the windows. The louver planned for sunlight is also designed in the respect of the whole module.

The elevation are applied to the louver system.

The piloti mass is connected to the western campus entrance from the courtyard of the education building.

1 T24 CLEAR PAIR GLASS
2 T3.0 ALUMINUM SHEET
3 T30 GRANITE SMOOTH TRIMMING
4 T5.0 ALUMINUM EXTRUDED BAR / ELECTROSTATIC POWDER COATING(FOR EXTERIOR)
5 INSTALLING HANDRAIL("B1" TYPE)
6 ALUMINUM GRILL
7 ALUMINUM EXTRUDED PLATE(60X450X2T)
8 WESTERN RED CEDAR
9 CEMENT MORTAR / WATER-BASED PAINT
10 CURTAIN BOX / T1.2 STEEL PLATE / ELEC-TROSTATIC POWDER COATING
11 T10 TEMPERED GLASS
12 IRON WOOD(18X89) / IRON WOOD(35X64-@500) / IRON WOOD BASEBOARD-(18X89X100@600) / DUST-PREVENTING PAINT / T70(MIN) PLAIN CONCRETE(#8-150X150 WIRE MESH) / W-3 WATER-PROOF / CONCRETE SURFACE FINISHING
13 INSTALLING HANDRAIL("B2" TYPE)
14 D-2 INSULATION
15 D-7 INSULATION
16 STEEL TROWEL FINISH / T70(MIN) PLAIN CONCRETE(#8-150X150 WIRE MESH) / T5 WATER-PROOFING PROTECTIVE PLATE / W-3 WATER-PROOF / CONCRETE SURFACE FINISHING

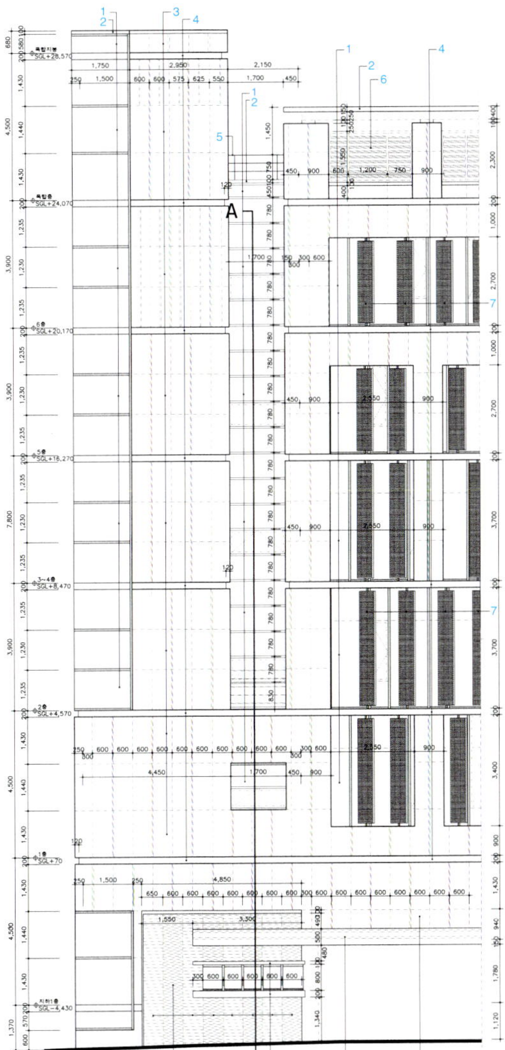

WEST WALL ELEVATION

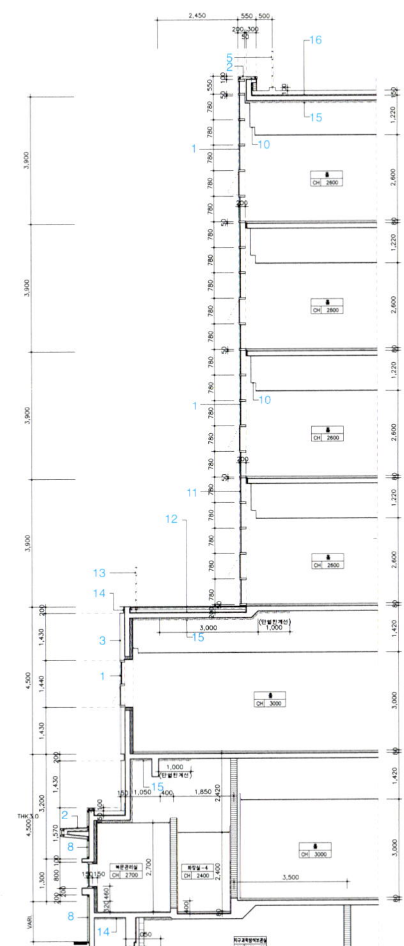

SECTION A

Elevation and piloti space viewed from existing education building

1 ALUMINUM GRILL
2 T30 GRANITE SMOOTH TRIMMING
3 T5.0 ALUMINUM EXTRUDED BAR / ELECTROSTATIC POWDER COATING(FOR EXTERIOR)
4 T24 CLEAR PAIR GLASS
5 T3.0 ALUMINUM SHEET
6 INSTALLING HANDRAIL("D" TYPE)
7 T10 TEMPERED GLASS
8 T5.0 ALUMINUM EXTRUDED BAR / ELECTROSTATIC POWDER COATING
9 T-200-100X200X12X12 / URETHANE PAINT
10 T24 OPAQUE PAIR GLASS
11 INSTALLING HANDRAIL("B1" TYPE)
12 STEEL TROWEL FINISH / T70(MIN) PLAIN CONCRETE(#8-150X150 WIRE MESH) / W-3 WATER-PROOFING / CONCRETE SURFACE FINISHING
13 D-7 INSULATION
14 CURTAIN BOX / T1.2 STEEL PLATE / ELECTROSTATIC POWDER COATING
15 INSTALLING HANDRAIL("B3" TYPE)
16 FIRE-PROOF PAINT(2 HOURS)
17 D-6 INSULATION
18 INSTALLING HANDRAIL("B" TYPE)
19 T0.8 ALUMINUM CEILING PANEL
20 D-2 INSULATION / T0.03 PR FILM / T60 LEVELING CONCRETE
21 W-4 WATER-PROOF / T3.0 ALUMINUM SHEET / T1.5 SELF ADHESIVE WATER-PROOF SHEET / T12 WATER-PROOF PLYWOOD 2PLY / □ -100X100X3.2T

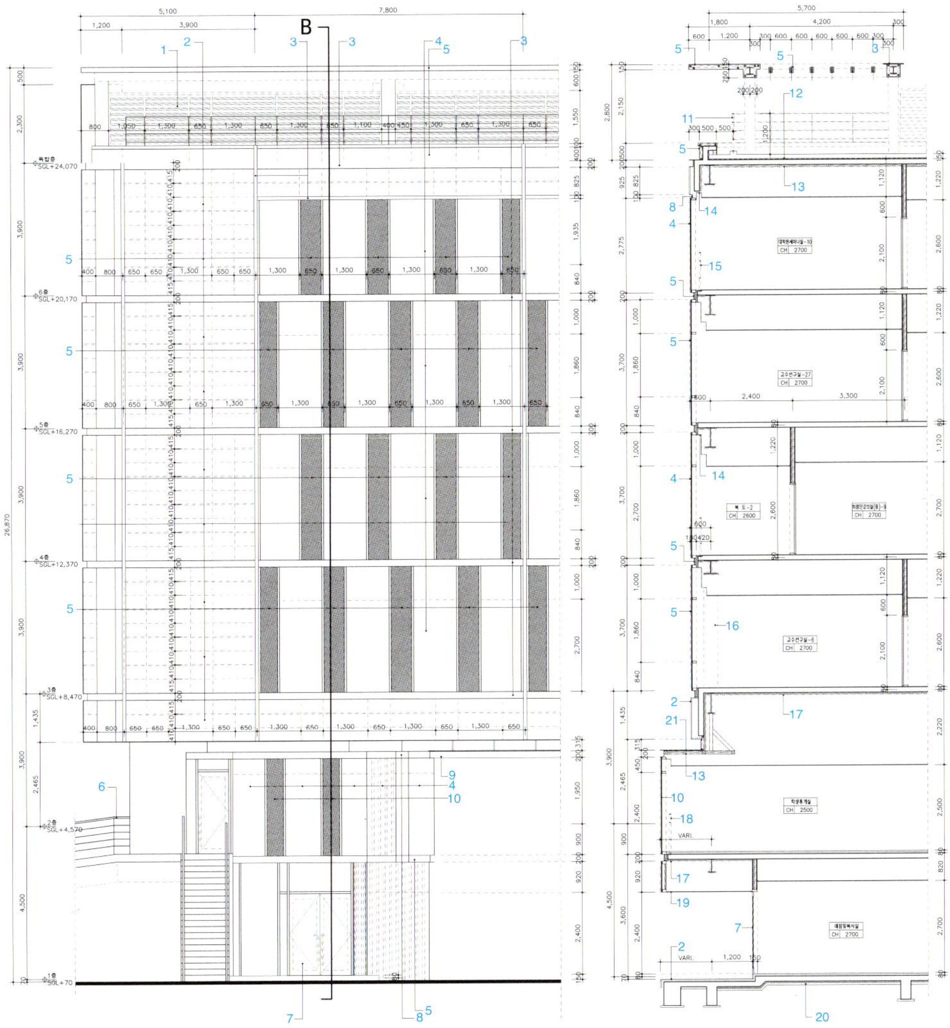

EAST WALL ELEVATION

SECTION B

Protruded snack bar and resting area

Piloti space

The sunken is a space for well-lighted of the underground space.

CONSTRUCTION

The building of the Education College, Ehwa Womans University has an affair related to a design, a structure and a construction different from the other buildings. At first, in the structural side, it is a matter to use an interlayer truss in order to make a Piloty to be connected from the internal central garden to the entrance of the western campus. Because this truss is exposed to the inside of the building, it may be a bit exceptional, considering a general emotion to mind an exposure of the structure. By removing four columns in the lower part of the Piloty through this steel truss we can acquire a sense of space not to feel a lower part of the building clearly.

The biggest characteristics of this building are the exterior design and the detailed construction. While most exterior of the building is made of the stone finishing, we would finish a design logic through a mixing use with other materials into it. As, first of all, the stone used in the north-western elevation is a part mainly exposed to the outside of the campus, we think that a university building to be opened to the local residents does not give an excessively heavy feeling. For this, the use of the aluminum bar between stones is proposed. We insert the aluminum bars every gaps between 6 stone panels and use wider aluminum bar than a vertical bar in every interlayer horizontally. So, the aluminum bars to be reflected with diverse shapes according to a strength of a light and glitter and the matte stones are combined and reduce a feeling of weight for the stone. An inconvenient thing is that a construction of the stone proposed as a drywall construction is changed into a semi dry type because of the construction cost. We show a shape of the stone laid on the aluminum bar, but we are sorry that owing to such a construction change, the aluminum and the stone are transmitted into a shape stuck at the concrete wall simply.

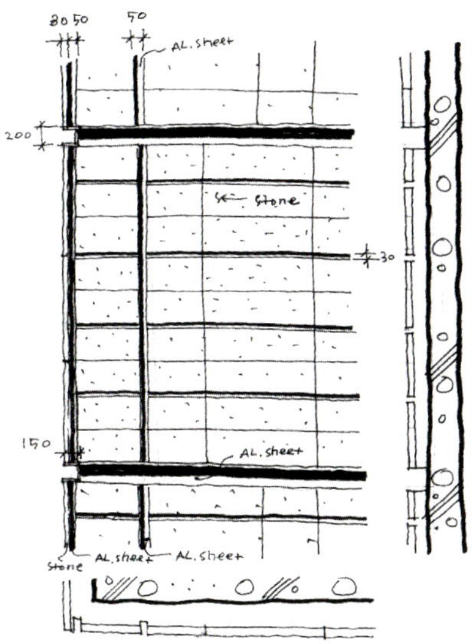

EXTERIOR FINISH DETAIL SKETCH

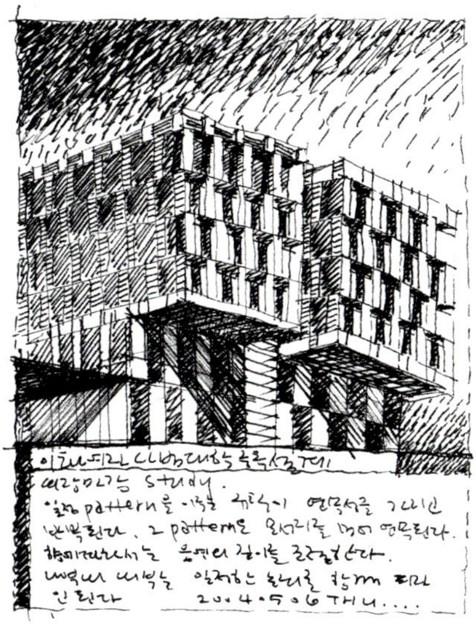

ELEVATION CONCEPT SKETCH

General view of the lobby

3RD FLOOR PLAN

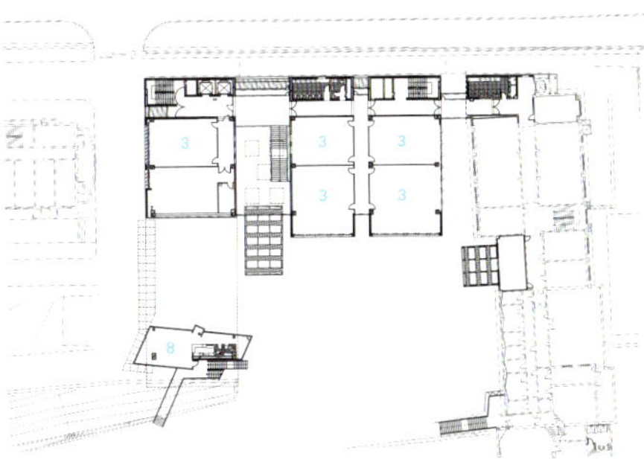

2ND FLOOR PLAN

1 MULTIMEDIA ROOM
2 SEMINAR ROOM
3 LECTURE ROOM
4 MULTIPURPOSE HALL
5 STUDY ROOM
6 MAIN ENTRANCE
7 SNACK BAR / COPY ROOM
8 RESTING ROOM
9 PRACTICAL ROOM
10 DIRECTOR'S ROOM
11 ADMINISTRATION OFFICE
12 OFFICE
13 PROFESSOR'S OFFICE
13 LABORATORY
14 STORAGE
15 MULTIPURPOSE HALL
16 PARKING LOT
17 ELECTRIC ROOM
18 EXISTING EDUCATION BUILDING
19 CORRIDOR
20 HALL
21 WIND-PROOF ROOM
22 LOUNGE

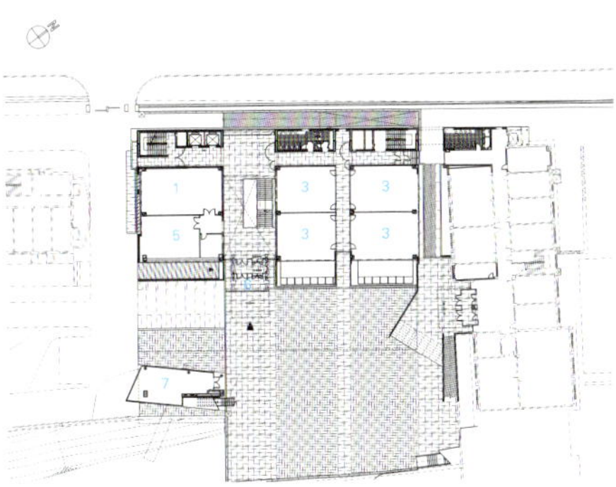

1ST FLOOR PLAN

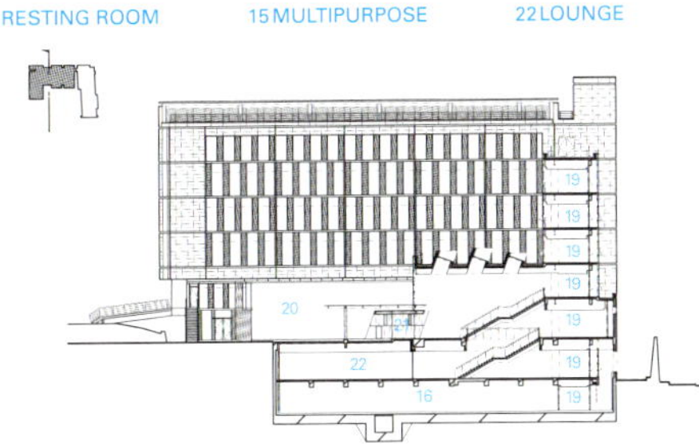

LONGITUDINAL SECTION

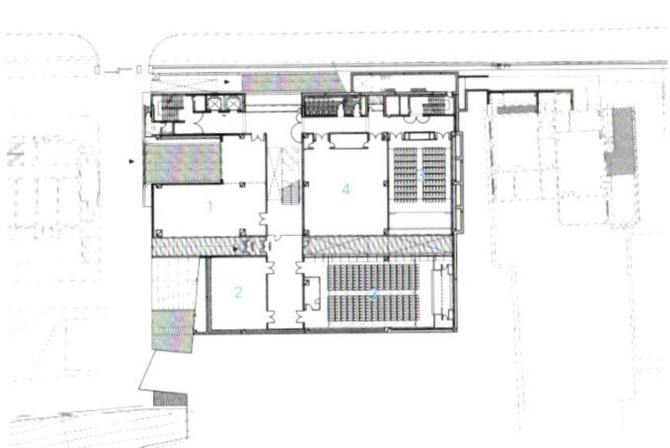

B1 FLOOR PLAN

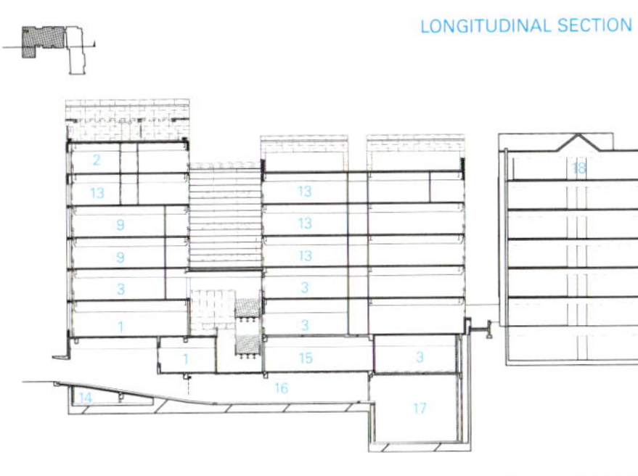

CROSS SECTION

KOREA UNIVERSITY HAESONG LAW LIBRARY

Location Anam-dong 5-ga, Seongbuk-gu, Seoul, Korea **Site Attribute** General Resident, Natural Scene **Site Area** 640,377m² **Building Area** 717.09m² **Total Floor Area** 3,670.20m² **Building Scope** B2~3F **Structure** RC, Steel **Exterior Finish** Titanium Zinc Sheet, Granite Burner Finishing, Color Low-E Pair Glass **Interior Finish** Exposed Concrete, Granite, Veneer Wood, Carpet Tile **Design Period** 2004.12~2005.11 **Construction Period** 2006.2~2007.3 **Architecture Design** Lee In-ho, Jung Min-hee, Kim Hyung-il | E.RAE Architects **Design Team** Jeong Min-hui, Jeong Sin-ae, Oh Dong-yeong **Interior Design** E.RAE Architects **Construction** Byeon In-geun, Kim Hyeong-tae | JAD Construction **Client** Korea University **Photographer** Kim Young-kwan(Offering from E.RAE Architects)

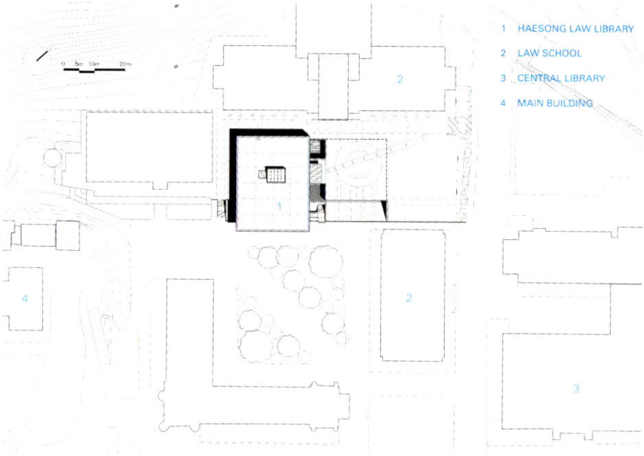

SITE PLAN

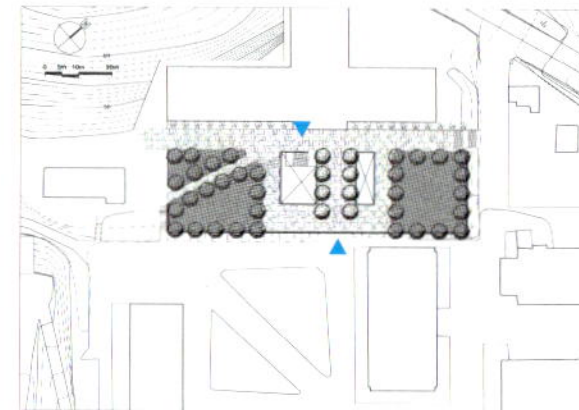

SITE PLAN ALT 1

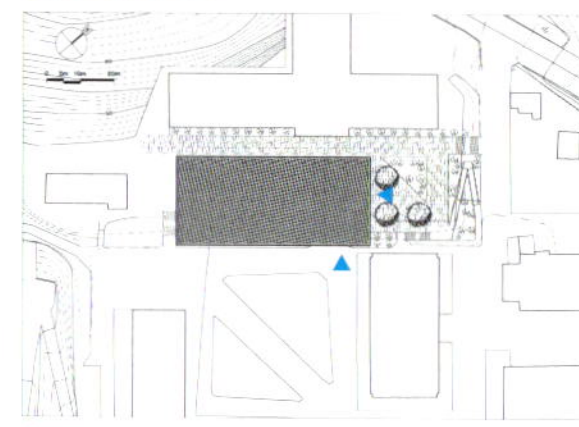

SITE PLAN ALT 2

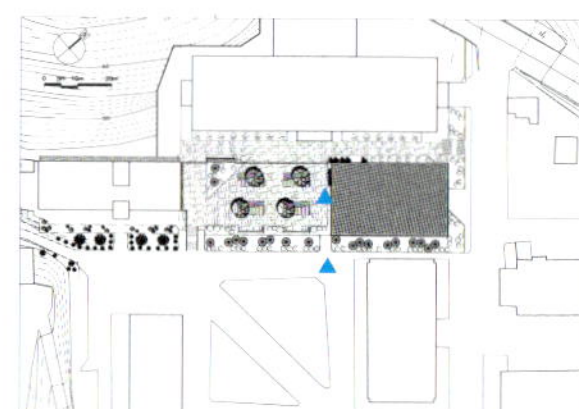

SITE PLAN ALT 3

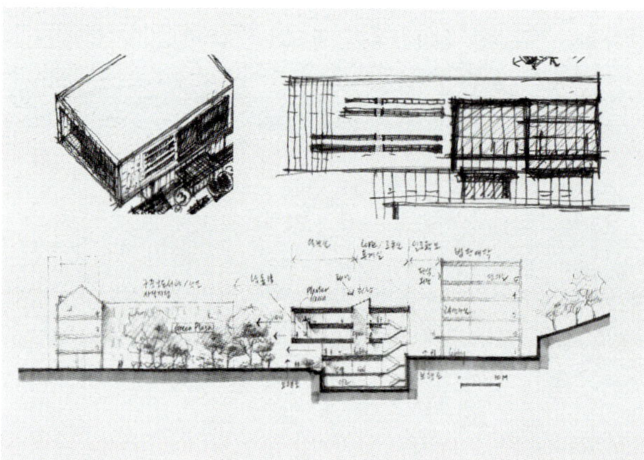

SKETCH

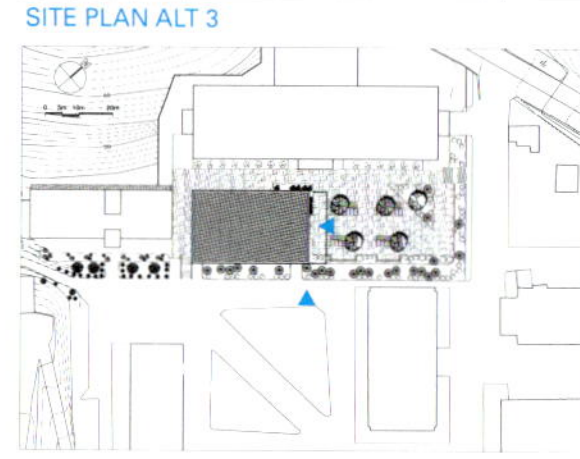

SITE PLAN ALT 4

The law library is shaped in a form which is inconsistent with the other structures of the school, but the main building with its pure and simple details is aimed at creating harmony in design along with the old central library's modern architectural style and designs. To preserve the existent buildings on the front and back sides of the library, the lecture rooms and reading rooms were put in the basement to minimize the mass of the building, and titanium zinc sheets similar to the color of granite with simple patterns were used as the main exterior surface. The natural rumple of the titanium zinc sheets harmonizes better with the rough texture granite than the refined plank stone finishing. The use of titanium zinc sheets as the surface material also reduce the fixed load of the 8.7m cantilever. The majority of the front sides of the first basement floor and ground floor on the south side have been processed with transparent glass to have a see-through and reflecting effects. The height of the building was decided by taking the height of the trees of the garden in the south into consideration, and as a result, the surrounding natural elements were brought into the projected reading rooms of the 2nd and 3rd floor. The 2-story high cantilever was projected towards the south side, and a reading room was situated here to open up the front side of the law school to its maximum capacity; furthermore, the landscape square and walking space of cantilever's lower sides were activated. The center of the building is open up to the third floor, thus the alternating space comes into view when walking up the stairs, and the ceiling of the same location allows natural light flowing into the building. The interior concrete walls and the steel frame material that support the projecting portion of the 2nd and 3rd floor have been naturally exposed. The spacial situation of the reading room has been made by prioritizing the importance of its resting function rather than its study function. The reading rooms in the open and bright lounge configuration were made by reflecting the study methods and attitudes of the rapidly changing new generation.

ㄱ-shaped simple mass

STUDY MODEL

General view of the main entrance

1 LEADING ROOM
2 NEWSPAPER ROOM
3 INFORMATION RETRIEVAL ROOM
4 OPEN-STACK LEADING ROOM
5 HALL
6 OFFICE
7 MACHINE ROOM
8 LOBBY
9 CASE STUDY ROOM

Side view of the main entrance

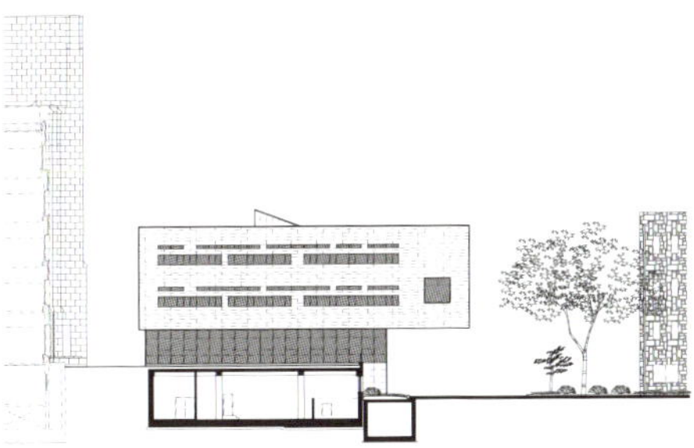

LEFT ELEVATION

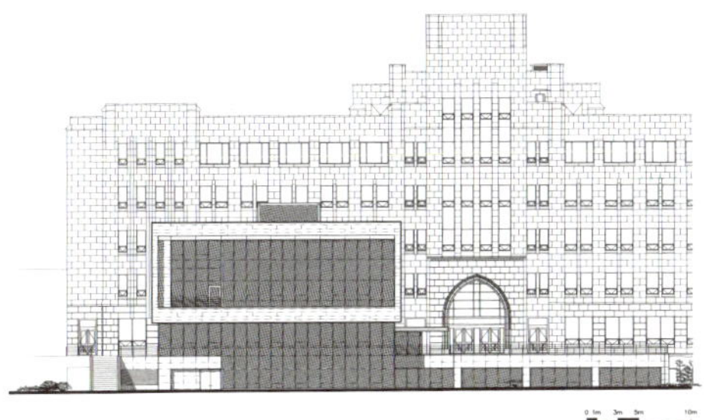

FRONT ELEVATION

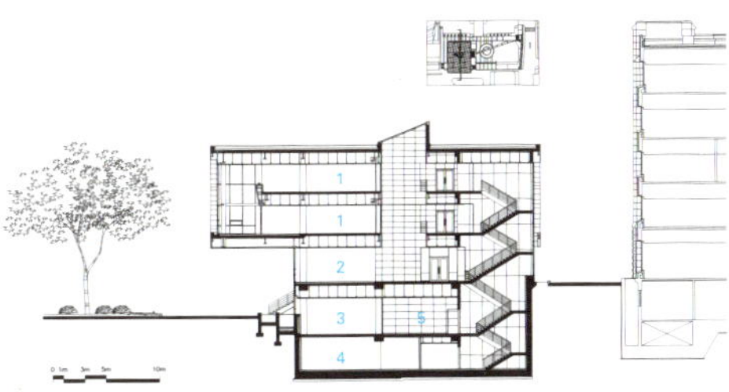

LONGITUDINAL SECTION

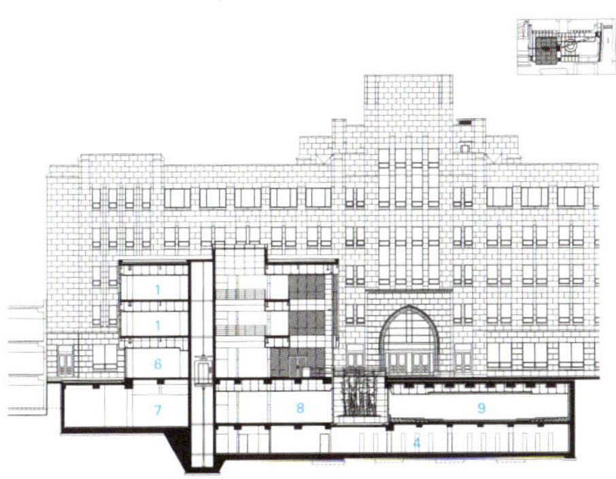

CROSS SECTION

The transparent glass allows an open view between the outdoor and the indoor.

The leading room viewed from the hall on the 3rd floor

The lobby on the 1st floor

The sunken on the B1 floor

The skylight viewed from the lobby

1 MAIN ENTRANCE
2 SUB ENTRANCE
3 LOBBY
4 MACHINE ROOM
5 CASE STUDY ROOM
6 LEADING ROOM
7 INFORMATION RETRIEVAL ROOM
8 RESTING AREA
9 SUNKEN
10 NEWSPAPER ROOM / RESTING AREA
11 SEMINAR ROOM
12 EXHIBITION GALLERY
13 OFFICE
14 HALL
15 STUDY ROOM
16 BALCONY

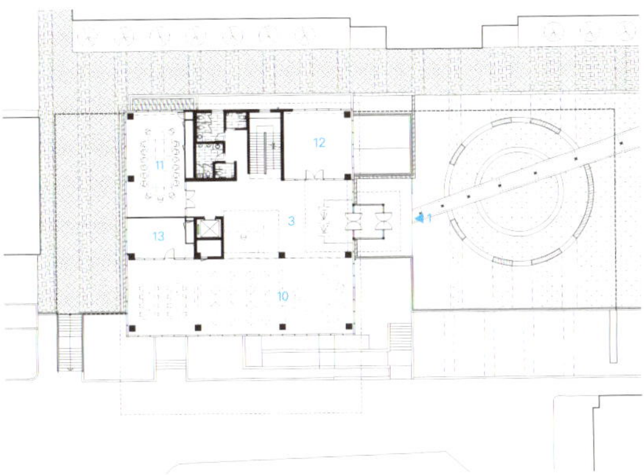

1ST FLOOR PLAN

3RD FLOOR PLAN

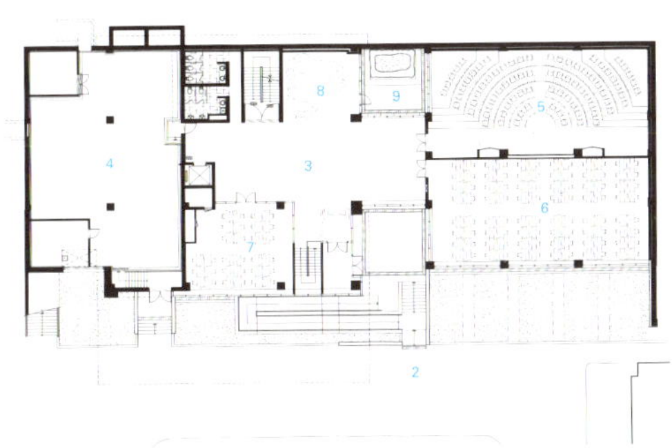

B1 FLOOR PLAN

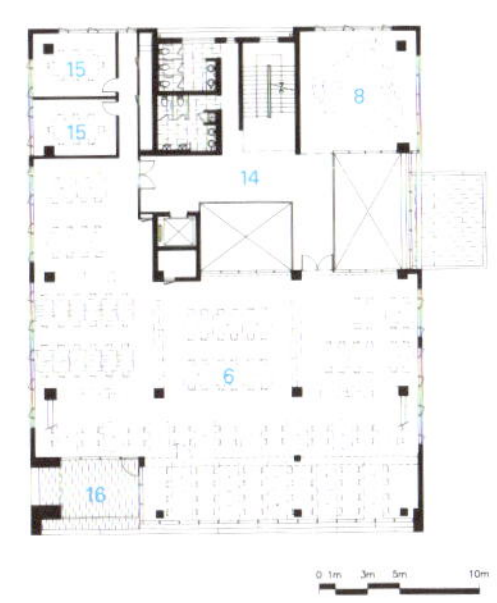

2ND FLOOR PLAN

View from the balcony on the 3rd floor

>> EDUCATION

JO MAN SIK MEMORIAL HALL & WESTMINSTER HALL

Location Sangdo-dong, Dongjak-gu, Seoul, Korea **Site Attribute** General Resident, General Aesthete **Site Area** 121,608.20m² **Building Area** 5,224.30m² **Total Floor Area** 31,862.59m² **Building Coverage Ratio** 25.60% **Floor Area Ratio** 105.54% **Building Scope** B2~7F **Structure** SRC **Exterior Finish** Granite Burner Flaring, Titanium Zinc Sheet, IPE, Aluminum Punching Plate **Design Period** 2003.9~2004.12 **Construction Period** 2005.4~2007.7 **Architecture Design** Lee Sung-kwan | HANUL Architects & Engineers Inc. **Design Team** Kim Hyun-jo, Jeoung Gye-baek, Uhm Duk-heum, Hun Hyun-sang, Kim Dae-hee, Cha Eun-kyoung, Lee Byoung-ho **Construction** SSANGYONG Engineering Construction Co., Ltd. **Client** Soongsil University **Photographer** Lee Ki-hwan

SITE PLAN

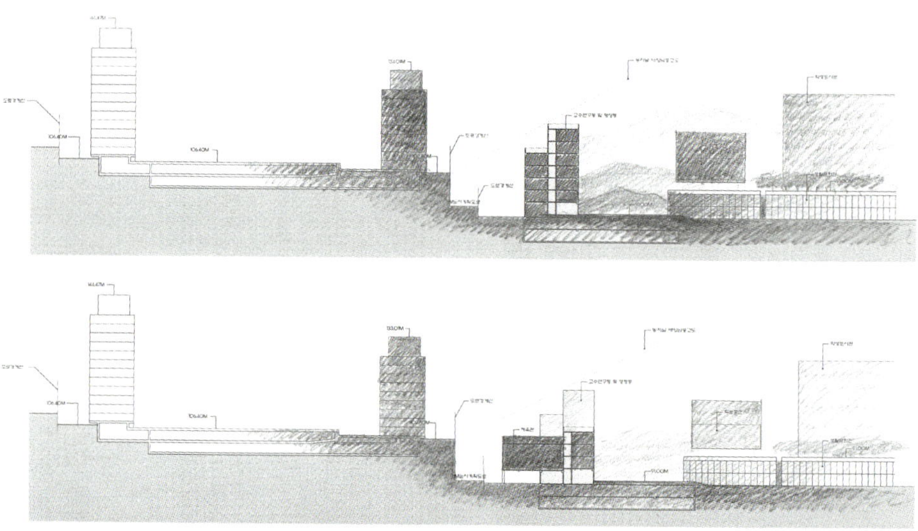

SITE SECTION

STUDY MODEL

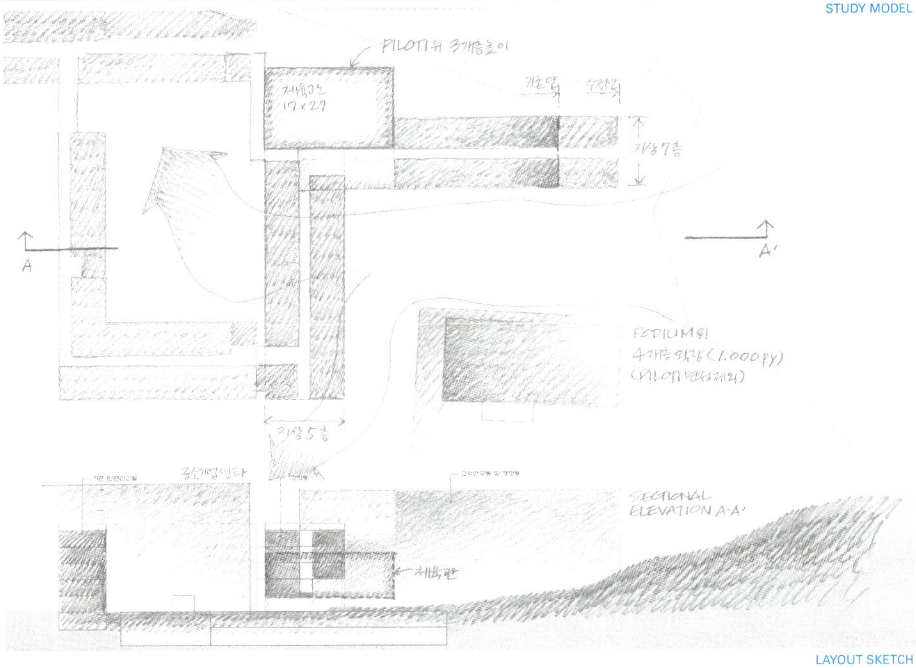

LAYOUT SKETCH

The construction of the Composite Lecture Center was in the plan that was promoted by the result of a prize contest in September 2002 for urgent tasks to improve outworn equipment in the campus and secure facilities. The plan was to withdraw the old three-story chapel that was located near Small and Medium Enterprise Center, Civilization Hall and the central library and newly build the "Composite Lecture Center(changed its name to Jo Man Sik Memorial Hall & Westminster Hall after completion of the construction work) " with seven stories above ground and two below that contains lecture rooms, offices for professors, graduate lecture rooms and laboratories and an indoor gymnasium. Soongsil University that stands in the lower edges of Mt. Seodal has potential energy for development as a green campus due to its geographical position between nearby urban areas and nature. In this view, the established old chapel that occupied an important geomorphological position didn't overcome topographical handicaps of the slope way in the process to induce Mt. Seodal into the campus and cut the connection between the campus and nature. To solve this problem, the basic concept of the new Composite Lecture Center was to "cure" the disconnected flow and "restore" it to a space with new spatial potential energy. The starting point of the Composite Lecture Center was here and it developed the concept of Green Spine that became a main structure of Soongsil University Master Plan(SSUMP), which won the contest later on.

The overall shape is "⊏ shape" and it was divided into two masses of " ¬-shaped " mass and " box-type " mass with piloti-processed bottom in the consideration of nearby buildings. Two bridges connect these masses; the bridge on the 3rd floor goes to the rooftop square of Welfare Hall that stands in front of the southern central library to raise accessibility to the central library and Welfare Hall and efficiency of the rooftop square, and the other bridge on the 5th floor secured accessibility and mobility between facilities. Also a new bridge was additionally built between Civilization Hall in the west and Small and Medium Enterprise Center in the north to raise efficiency in transfer and study activities through the organic connection between buildings and leave possibility to expand the building if necessary.

The external space between "¬-shaped " mass and " box-shaped " mass was planed to induce the flow of Mt. Seodal into the campus. The upper part on inclined land became a resting space with wooden terraced stand to draw the flow of Mt. Seodal into the campus and the lower part contains a multipurpose indoor gymnasium including a fitness center and a jogging track.

Text by HANUL Architects & Engineers Inc.

Main view of the composite lecture center

Night view

Facade of the westminster hall

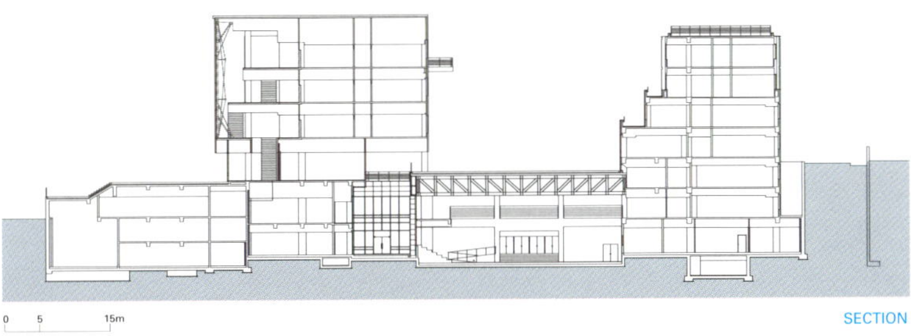

SECTION

A resting space with wooden terraced stand

The wooden terrace encircling the buildings

The resting open space

1 MULTIPURPOSE GYMNASIUM
2 PHYSICAL TRAINING ROOM
3 HALL
4 LECTURE ROOM
5 PARKING LAMP
6 MANAGEMENT ROOM
7 RESTROOM

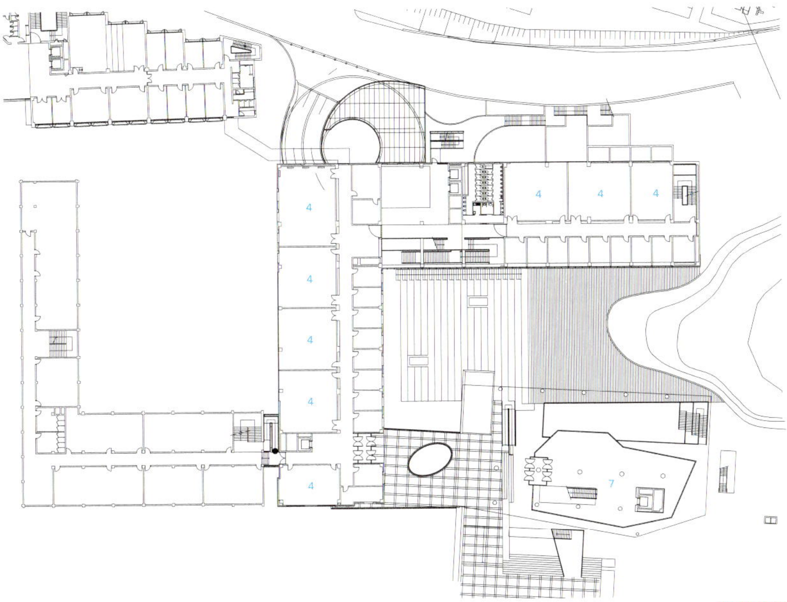

3RD FLOOR PLAN

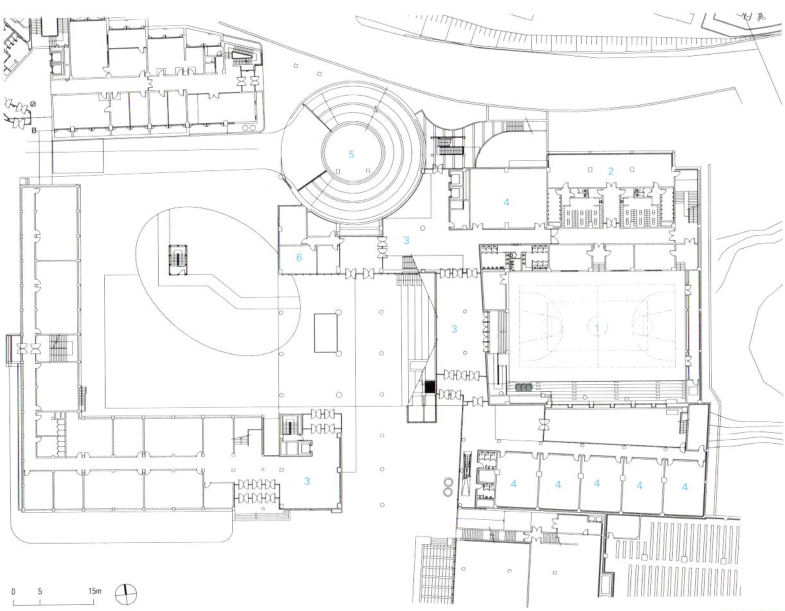

1ST FLOOR PLAN

Open Stairway

Inner space art wall

Open Stairway and the resting space

>> EDUCATION

HEUKBIT YOUTH CULTURE CENTER

© Park Young-chae

© Park Young-chae

Location Gohan-ri, Gohan-eup, Jeongseon-gun, Gangwon-do, Korea **Site Attribute** General Resident, Nature Scene **Site Area** 1,674m² **Building Area** 851.31m² **Total Floor Area** 1,647.37m² **Building Scope** B1~3F **Design Period** 2004.11~2005.11 **Construction Period** 2005.12~2007.3 **Architecture Design** Kwak Jae-hwan + Lee Young-ik | DONGWOO Architects & Consultants **Photographer** Park Young-chae, DONGWOO Architects & Consultants(Except Otherwise Indicated)

STUDY MODEL

PROLOGUE - GOHANEUP Gohan-eup which had grown rapidly as a new mining town in the past faced a crisis because of the policy of mining industry rationalization carried out late 1980's. In order to support the mining town, by establishing the special act on support for abandoned mine area in 1995, the Gangwon Land was built in Gohan-eup, and the Gangwon Land changed in a short period the image of Gohan, an area of an abandoned mine. Since then, although Gohan has an affluence in the economical side, the prosperity and decline of the town for a short time does not have a spaceous consideration to raise the 'quality of life' for the residents. In particular, a space which the teenagers who are the future of the town can spend a beneficial time is none.

HEUKBIT YOUTH CULTURE CENTER-ONE HEART FOR ALL In Gohan Eup, there was the 'Heukbit Youth Culture Center' which the village cathedral operated as a meeting place for teenagers. 'Heukbit' implies a mining town but rather, symbolizes a night sky and contains an adult's mind that the children who come here become shining stars of a night sky. Because the building which this youth center rent was destroyed, it had held its place in the cathedral for some time. Hereby, the local officials and civilians including the Gohan cathedral, the Wonju District of Catholic, Jeongsun County Hall and the local civil organization, the promotion headquarter for campaign to revive Gohan·Sabuk·Nammyeon collect their mind to make a decent building. In spring, 2004, we decided a design plan through about 1 year-work and held a construction ceremony on Mar 19 next year. After the construction ceremony, the agony continued. It should be a better house for the youth, further a culture space for the local residents and again further we would make a spiritual & cultural central space of the local society. In the time of agonizing, the architect, Jae-hwan Kwak, actively to respond the meaning of the youth center participated in the work together. Priest, Woo Gap Lee, Gohan Cathedral, to hold an architecture expressed his expectation for the building with a word 'the common is a crime' to the new youth center. This building is the house which the souls of the miners who dug coals in the dark underground and their children made and they are the people who realized an existence of the light the most urgently. Maybe, it means an aspiration toward a brightly shining house to express the most symbolically a change in the area of the abandoned mine like Sabuk and Nammyeon around the Gohan. The building finished on March, 2007 was made as the bigger, more open and brighter house than an original plan.

ARCHITECT, BRIGHT AND WARM LANDMARK The Heukbit Youth Culture Center placed on the hill facing the south to look down the Gohan Station, Taebaek line become new landmark in Gohan. In common, an architecture might be a process of filling. However, it is kept in mind that the Heukbit Youth center is a learning center and a playground for the children and the teenagers

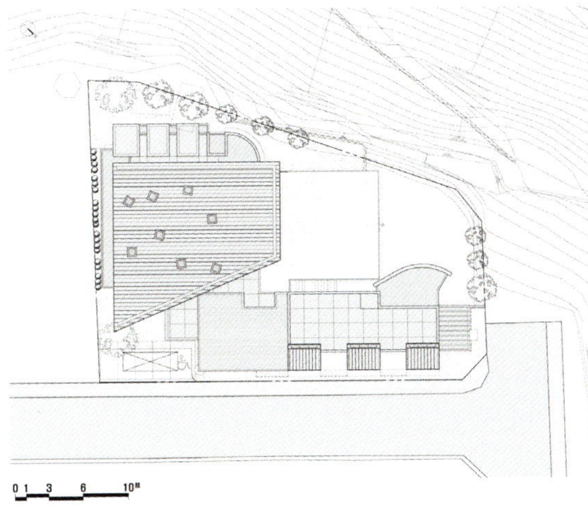

SITE PLAN

and a meeting place for the local residents. For this, in order to make a space with full sunshine, which is convenient, familiar and vivid to all the people from children to the elder, I thought how to make empty. The outer space has a two courts with different level and both courts are a place to view the Gohan area at a glance. The high court, the level of the first floor, is raised for its usage by making the deck and connecting with the multipurpose theater without faulting. The low court of the level of the first basement is made of a bright and vivid space with full sunshine indoor and outdoor by making a garden and connecting the learning center finished with glass in whole wall. The outdoor deck is directly connected with the performance hall and possible to actively utilize the inner and outer spaces. The outdoor court to induce a nature's graceful scenery into the indoor space is an outdoor learning space connected to the indoor learning center and a safe playing space. It is a town to have no cultural space for the local society as much as a wasteland of culture and the most urgently necessary thing for it is a space with diverse usages. Hereby, it is made of wide performance with facilities. In the ground, the wooden floors are spread so that sometimes it is used as a gymnasium and holds a party for a village. For more usage, it s connected with the outdoor deck. No faulting makes children run and the elder difficult for movement come and go out indoor and outdoor without inconvenience. Indoor spaces with diverse sizes are arranged around the outdoor deck and the outdoor court and become a place with a bright and warm energy full always. All the rooms are considered for the multipurpose usages and the pathways to connect the rooms themselves are a playing spaces to run, roll over and talk. There is no closed place. In the inside, we can see the outside and in the outside, we can the inside. The inside and the outside are one. Wherever staying and wherever going, all the things become one.

EPILOGUE - HOUSE TO CONTAIN ALL PEOPLE'S DREAM The real meaning of the Heukbit Learning Center and the Heukbit Youth Culture Center lies in the process for making a building to be contained in the other side rather than a configuration of the architecture seen in the eyes. Even in the difficulties to be born as a mining town and grow, driven in the crossroads of the life and death owing to abolishing a mine and go through a rising to the amusement town, the house built as the earnest hearts of the people not to release their dreams toward future. Therefore, the Heukbit Learning Center, although an outer appearance is a small house, becomes a very big house to contain a big picture which many people draw inside. The people of Gohan and the people of Jeongsun County come to have a source of pride. Not a pride to have a nice building, they have a pride to do anything if they collect their minds and gather their forces one another in spite of small force.

The garden on the B1 floor

Night view © Park Young-chae

View from the outdoor deck on the 1st floor © Park Young-chae

The outer wall viewed from the outdoor deck

Night view of the outer wall viewed from the outdoor deck © Park Young-chae

The outdoor deck is connected to the performance hall.

1 OUTDOOR DECK
2 PERFORMANCE HALL

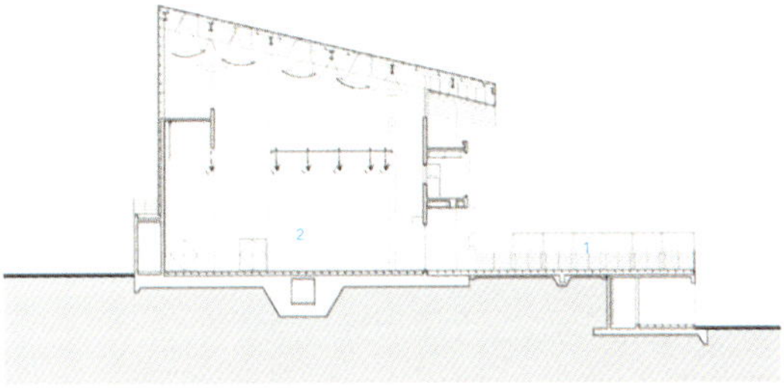

SECTION

General view of the performance hall

The ceiling detail of the performance hall

The outdoor deck viewed from the performance hall

General view of the learning room © Park Young-chae

Void staircase

General view of the staircase

1 GARDEN
2 OUTDOOR DECK
3 PERFORMANCE HALL

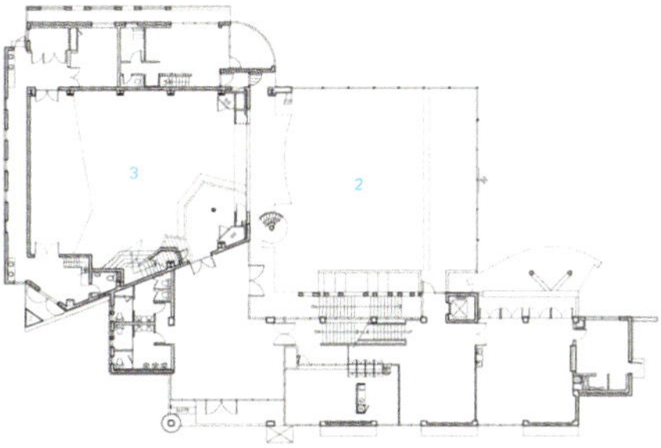

1ST FLOOR PLAN

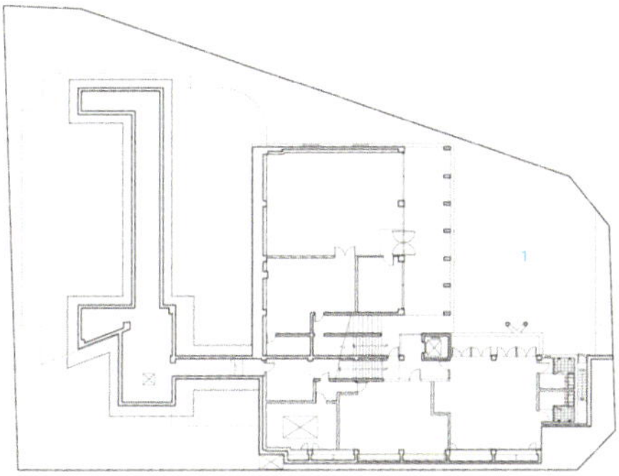

B1 FLOOR PLAN

The staircase leading to the outdoor deck

View from the staircase

INCHEON ENGLISH VILLAGE(ICEV)

Location Dangha-dong, Seo-gu, Inchon, Korea **Site Area** 9,494m^2 **Floor Area** Main Building 3,987m^2 (1F_968.2m^2, 2F_958.9m^2, 3F_951.3m^2) / Service Building 1,953m^2(1F_485.8m^2, 2F_559.5m^2) / Dormitory 2,989m^2 (1F_728m^2, 2F_455m^2, 34F_391m^2, 5F_351m^2) **Building Scope** Main Building_4F / Service Building_3F / Dormitory_5F **Exterior Finish** Punched Metal, Metal on Paint and Heat-Dry, Wood on Oil Stain, Special Water-Soluble Paint **Interior Finish** Vinyl Paint, Water Paint, Sheet, Plastic Tile, Carpet Tile, Chip Board **Design Period** 2005.10~2005.12 **Construction Period** 2005.12~2006.2 **Design** Kim Taek-soo | Ver-TEX Design **Design Team** Seo Kyoung-il, Min Hyo-kyoung, Han So-young, Mathew Uselman, Kim Tae-gyum, Jung Hae-jin, Kim Hyun-joo **Client** Lee Woo-young | Palbong Industrial Eeducation Inc. **Photographer** Yum Seung-hoon

General view of Inchon English Village

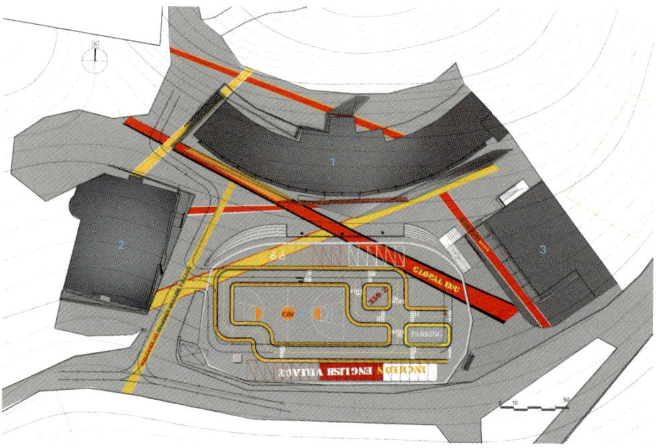

1 DREAMER'S HALL(MAIN BUILDING)
2 SERVICE BUILDING
3 DORMITORY
4 PLAY GROUND

SITE PLAN

STUDY MODEL

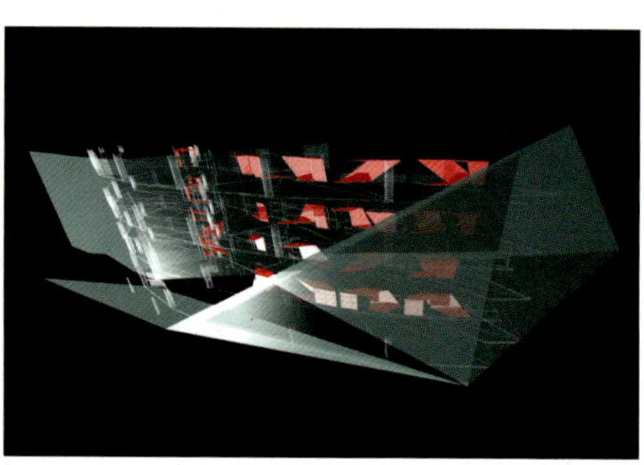

STRUCTURAL DIAGRAM

Entrance, dreamer's hall

STRUCTURE DEVELOPMENT

The double - layered screen

It is designed to let the children understand the given spaces.

The remodelling project of Incheon English Village includes the interiors, facades, the spaces between the buildings and the entrances of the Main Building (four floors), the Service Building (three floors) and the Dormitory (five floors). The main ideas for remodelling the facades starts from imagining new forms derived from the basic structures of the three existing buildings. The space is programmed on the fact that the main users are the elementary school students who want to learn English as a second language. It is designed to let the children understand the given spaces in fresh ways by interconnecting and expanding their perception of things with the space in an organic way. In order to enable the interplay between the children and the space, the idea of the design started with 'the continuity of origami.' The process of making various shapes out of a single sheet of paper develops into creation, assembly, organic composition, colours and spatial composition, acting as a stimulant to the children s mind. The process of folding, piecing together and fixing by the hands of the children becomes a new approach to designing the Main Building and its external facade. The facade of the Main Building is developed by covering the existing cylindrical building with fifteen pieces created by a process of successive variation. Each of the pieces is in close relationship of co-dependence both in structure and in shape, maintaining a balance of spatial power. The successive pieces of the facade call for a movement of inertia that would not stop, which then spins off the canopies, passages, landmarks and the intermediate spaces from the Main Building to the Service Building and to the Dormitory. Unlike the wet fixing method that begins with destroying the existing building, the structure of the perforated metal screens bracing the facade symbolises the emergence of preservation that is permeated in the entire project, a symbiosis that recognises and allows the past and the present and the old and the new. The double-layered screen fabric offers an ever-changing reflective surface, creating dynamic changes, allowing and refusing to blend with the existing structure continuously. An organic shape allowing and accepting free flow of energy rather than regulating and concluding it -- and the extraordinary flow of space it creates -- ceaselessly transforms only when it is harmonised with light and shadow. Text by Kim Taek -soo

The corridor, dreamer's hall

Lobby, dreamer's hall

Stair case

Life experience, airplane and car

Cooking studio

Library

>> EDUCATION

GUMI MUNICIPAL BONGGOK LIBRARY

Location Bonggok-dong, Gumi-si, Gyeongsangbuk-do, Korea **Site Attribute** Urban Neighborhood Park **Site Area** 11,837.20m² **Building Area** 1,539.13m² **Total Floor Area** 4,825.97m² **Landscape Area** 7,398.25m² **Building Coverage Ratio** 13.00% **Floor Area Ratio** 34.66% **Building Scope** B1~3F **Structure** Steel, RC **Exterior Finish** T12 Tempered Glass, T18 Low-E Pair Glass, T30 Granite Burner Flaring & Rubbing **Interior Finish** Granite Rubbing, T2.5 Non-Conductive Vinyl Tile, Acryl Paint, T12 Asbestos Free Sound-Absorbing Tex **Design Period** 2004.12~2005.6 **Construction Period** 2005.12~2007.7 **Architecture Design** Son Chang-ho | GROUP ONE Architects & Consulting + Kim Seok-man | Kumoh National Institute of Technology **Design Team** Jo Jae-ung, Kang Ki-jin, Kim Chang-hyeok, Mon Seong-in **Interior Design** GROUP ONE Architects & Consulting **Construction** CHEONGTO Construction **Client** Gumi City Hall **Photographer** Lee Ki-hwan

General view from the south-east

1 LIBRARY
2 CENTRAL PLAZA
3 COURTYARD
4 PARKING LOT
5 POND
6 PLAZA
7 STAGE
8 GARDEN
9 FITNESS TRAINING AREA
10 BASKETBALL COURT
11 MULTIPURPOSE GROUND
12 8M ROAD

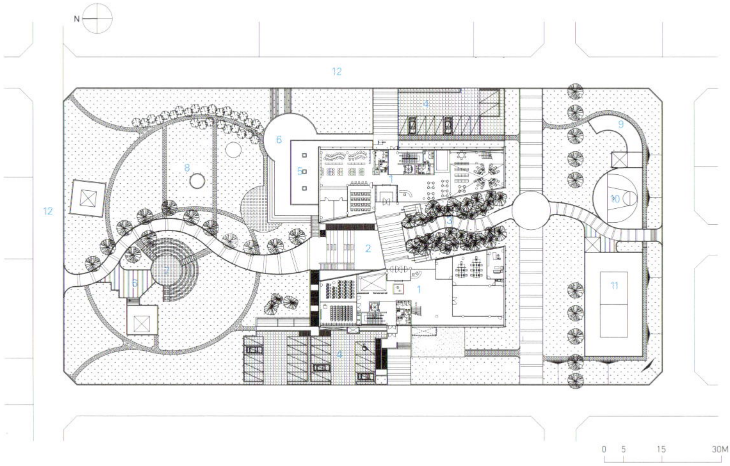

SITE PLAN

Night view from the south

SITE In the geometric 1:2(L:H) proportion of the rectangular site, the bridge that connects the two separate buildings placed respectively in the South and the North by forming an axis and fragmenting the mass. The site was designed in H shape to integrate a variety of functions and features of the library. The bridge, built in traditional style, emphasizes the basic configuration of the mass and functions of the passage. By arranging the long pine forest in the empty spaces between building and building, each building can have a better shading and view. The road which penetrates the east-west axis of the site links the adjoining roads with the blocks while the other road linking the south to the north is stretched to the lane between the History court and the Play court and to the pine forest.

FLOOR For both of separation and integration of the functions of a library, the library was designed in H shape which can organically connect every zone. Such separation and integration are achieved by a bridge which connects the two buildings to the core.

FACADE The eastern facade materializes a pine tree which is the trace of the original ground. The facade is formed of fragmented sections and vertical columns which materialized a great mixture of irregularity and regularity. The mass, finished with glasses, emphasizes the concepts of the library such as openness and clearness. And the sections which were finished with granite represent concise and moderate images.

SECTION The sections were built in efficient way, utilizing the difference of height with the adjoining road. Between the two buildings, a small arena was placed. The western section was designed to filter the strong sun rays, and the double-story study room built as a matrix in the children's library excites children's curiosity.

The transparent glass allows an open view between the interior and the exterior

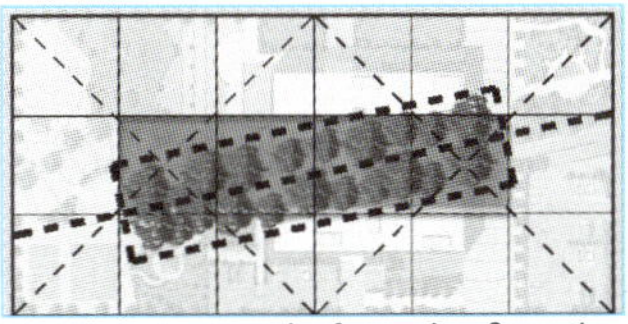

site formation & vestige

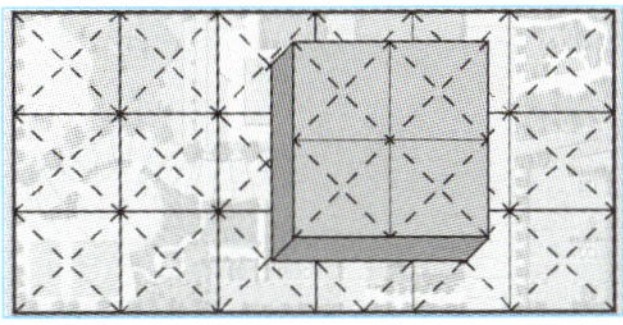

mass formation

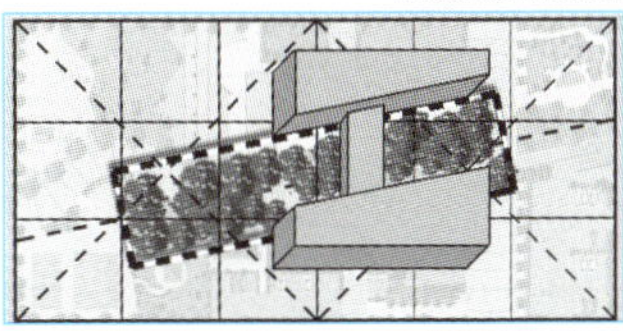

insert vestige

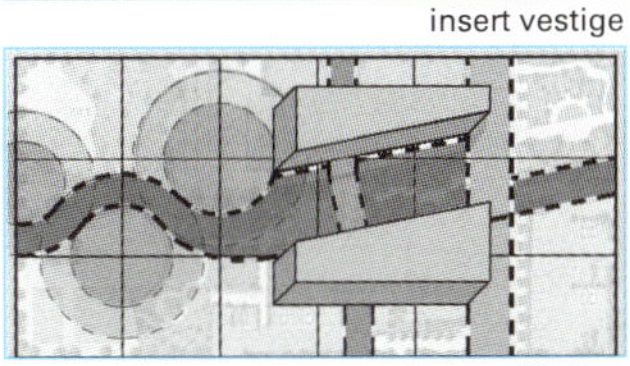

road reconstitution

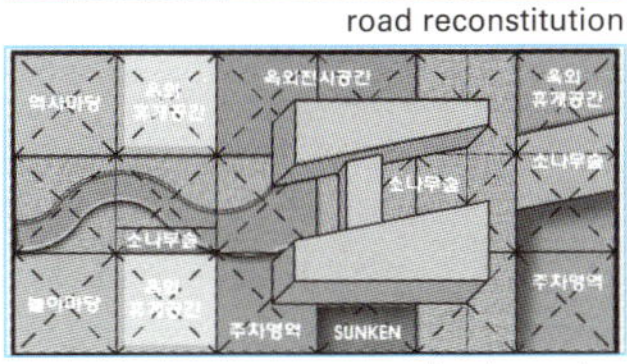

land use plan

SITE PROCESS

Patterns of pine on the elevation

pine forest

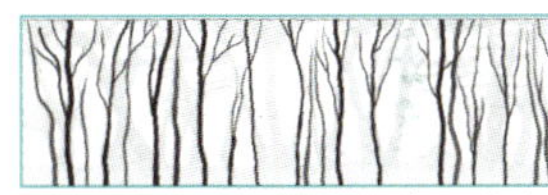

background & picture(regularity)

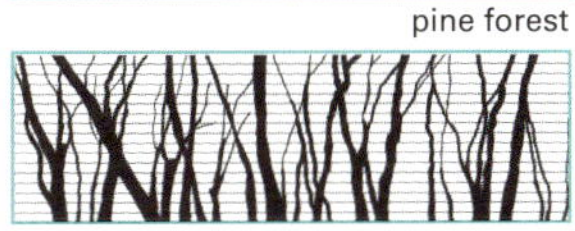

patterning

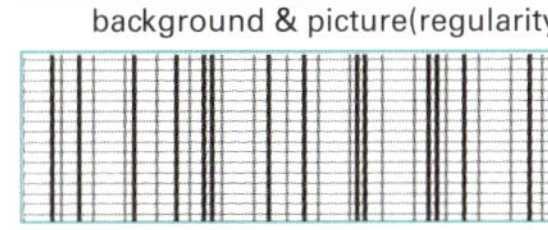

patterning

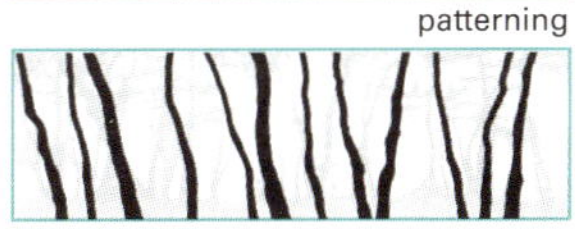

background & picture(irregularity)

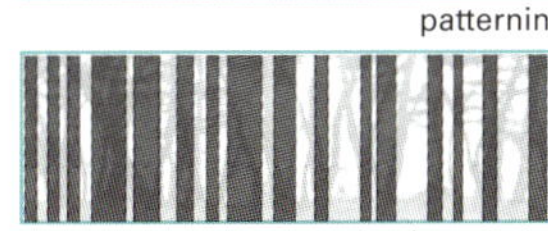

reconstitution(regularity)

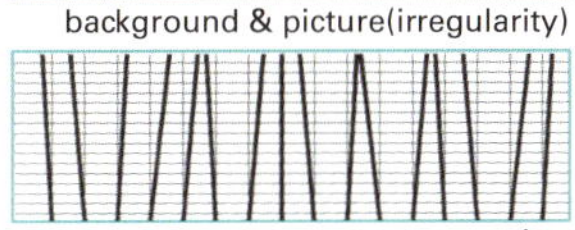

patterning

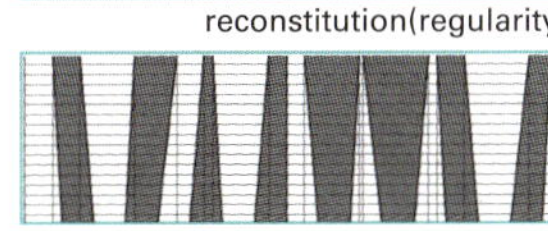

reconstitution(irregularity)

ELEVATION PROCESS

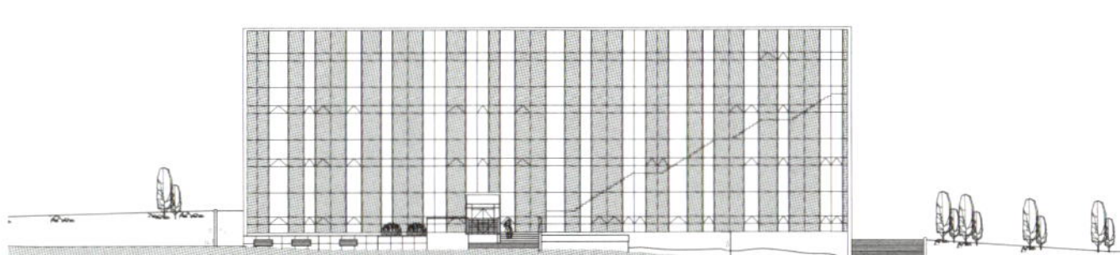

WEST ELEVATION

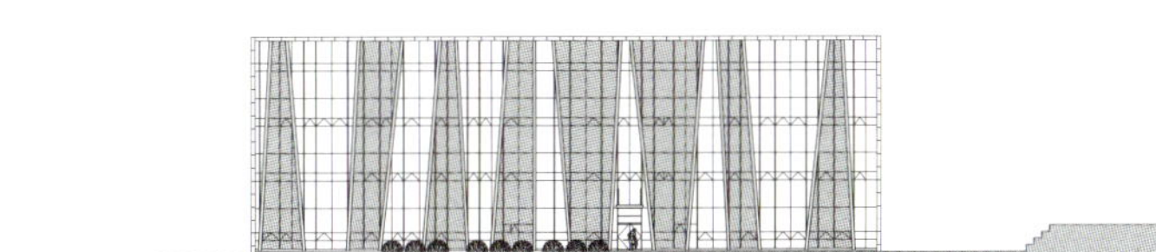

EAST ELEVATION

Interior of the eastern library

1 HALL
2 LECTURE ROOM
3 EDUCATION OFFICE
4 OFFICE
5 DIRECTOR'S ROOM
6 COMMUNICATIONS ROOM
7 LEADING ROOM
8 ORALLY NARRATED FAIRYTALE ROOM
9 STUDY ROOM
10 DATA ROOM
11 TOILET
12 LIBRARY

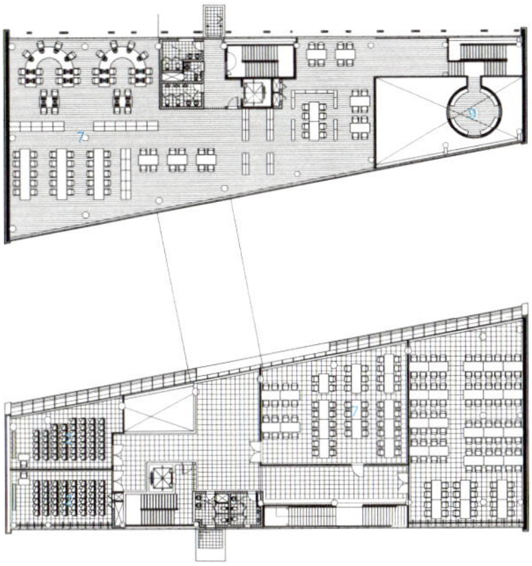

2ND FLOOR PLAN

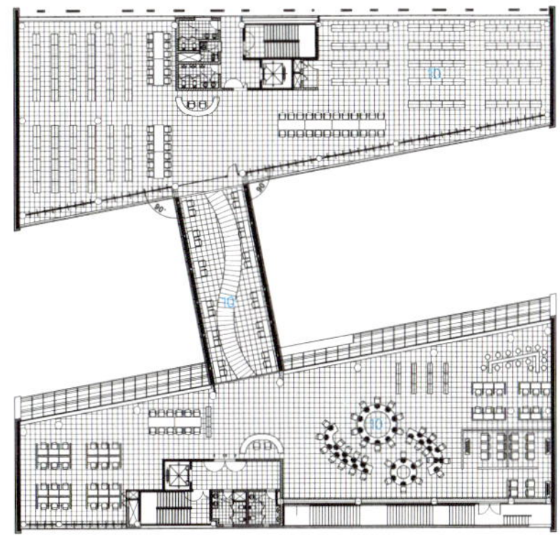

3RD FLOOR PLAN

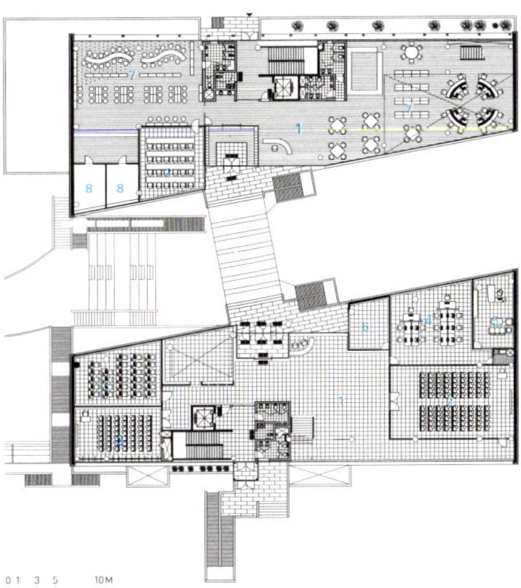

1ST FLOOR PLAN

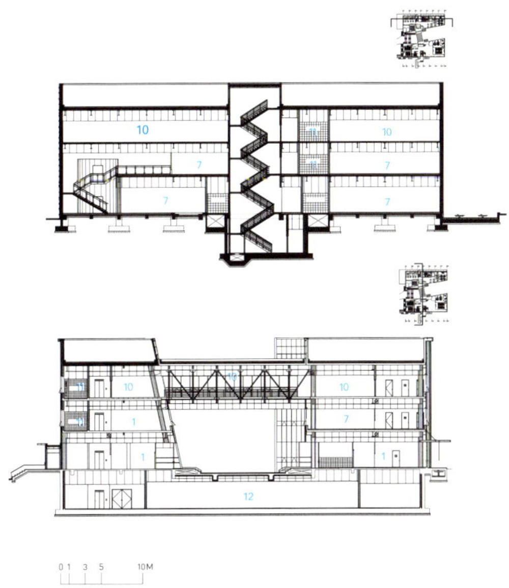

SECTION

JUYEOP CHILD LIBRARY

Location Juyeop-dong, Ilsan-gu, Goyang-si, Gyeonggi-do, Korea **Site Attribute** General Resident, District Unit Plan **Site Area** 1,500.00m^2 **Building Area** 748.33m^2 **Total Floor Area** 2,049.78m^2 **Landscape Area** 231.07m^2 **Building Coverage Ratio** 49.89% **Floor Area Ratio** 116.62% **Building Scope** B1~3F **Structure** RC **Exterior Finish** T30 Lime Stone(Red Color), T0.7 Titanium Zinc Sheet, Attaching Tiles(White Color+Red Color), Exposed Concrete, T16 Color Pair Glass **Interior Finish** Floor_Wood Flooring, Oil Stain on Lumber / Wall_N Noble Board, Copenhagen Rib / Ceiling_Sound-Absorbing Tex, Expanded Metal **Design Period** 2005.6.10~2005.9.7 **Construction Period** 2005.12.7~2006.12.21 **Architecture Design** Choi Dong-ho | Studio MADANGSORI **Design Team** Min Byeong-ho **Construction** SAEMIRR Construction & Engineering **Client** Mayor of Goyang-si **Photographer** Lee Ki-hwan

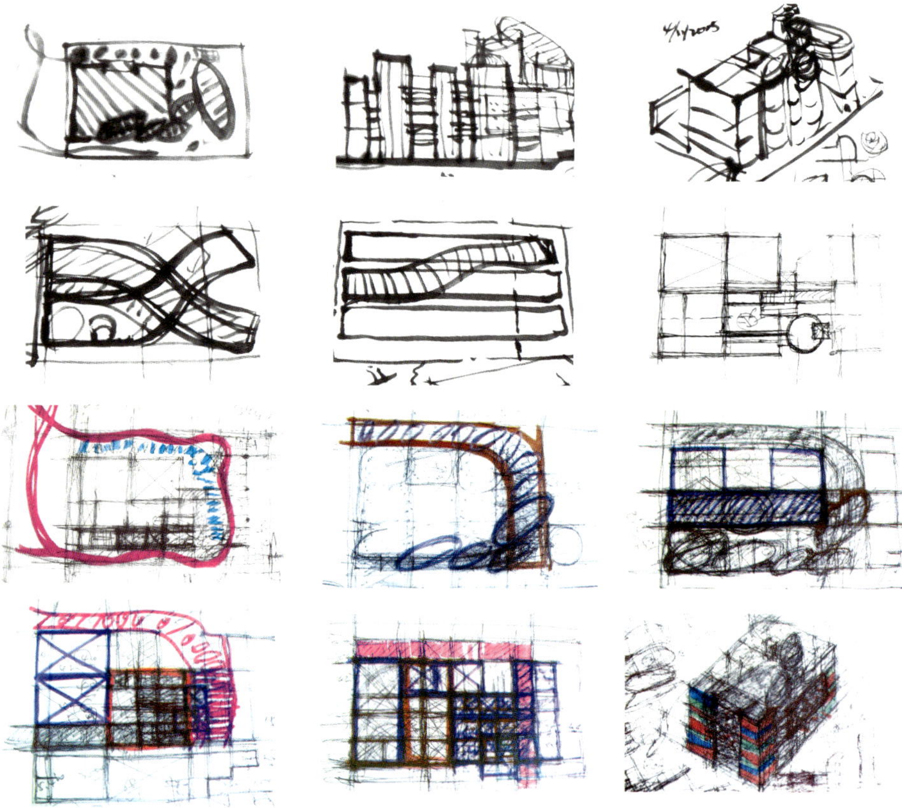

SKETCH

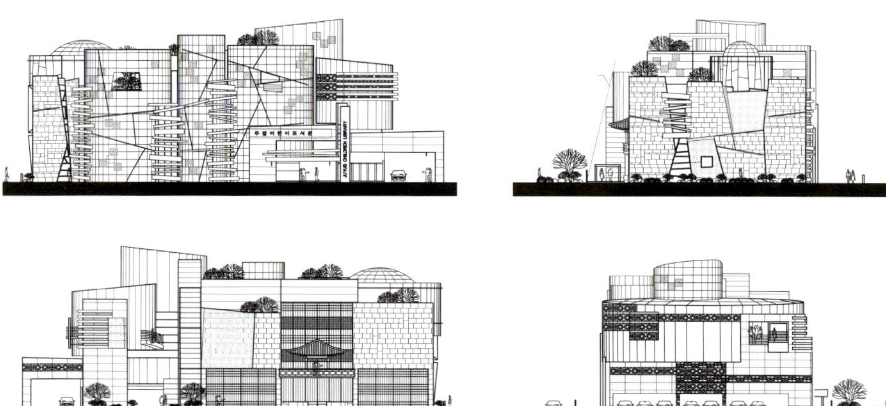

ELEVATION

1 MAIN ENTRANCE
2 SUB ENTRANCE
3 15M ROAD FOR WALKER
4 30M ROAD
5 ENTRANCE FOR CAR

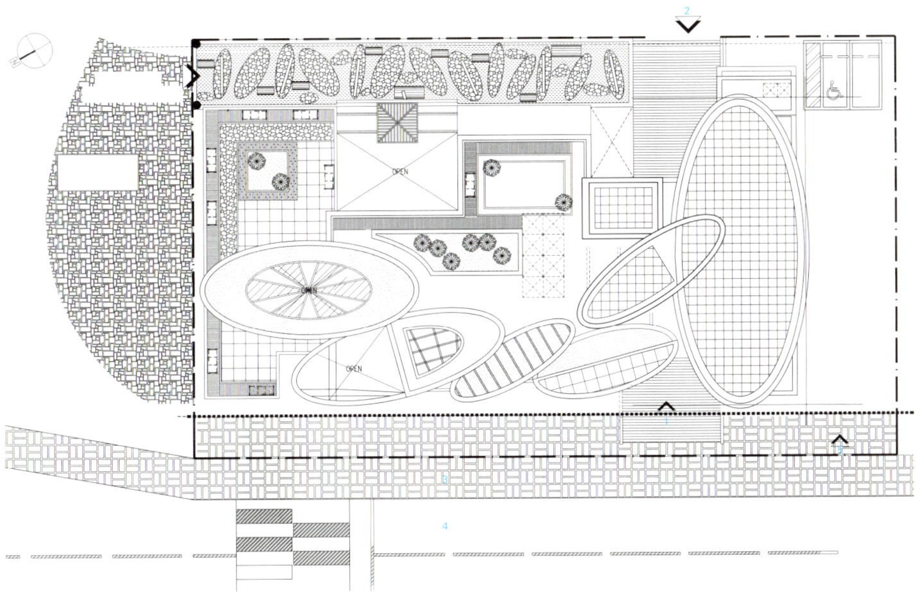

SITE PLAN

CHILD LIBRARY AND MANDAERU My heart beats as I approach Byunsanseowon. It is because I know that I can see such great architecture like Mandaeru there. I believe that traditional architectures are not to be seen but to be felt. I think the loft was made based on the Korean identity though it looks global on the other hand. I focused on materializing the core ideology of the 'touch' of Mandaeru through the child library. Thus, the library where children can read books standing on bare feet started.

"Juyeop Child Library" project is carried out based on several aesthetic concepts. As Hegel defined through the comparison of a tree and a seed that "a seed is a concept while a tree is reality", a concept is "realistic and simultaneous, and an ideological integration". Therefore, a child library should be designed for 'children' in an aesthetic view.

When you enter the reading room in the child library, you can face the inner court and the main floored room first of all. The floored room is opened to the court and the trees stand in a row like a wide screen while a stream from Nakdonggang river flows along the court.

The floored room for providing children with a space for reading and researching was constructed in imitation of Mandaeru. Even the floor pattern looks similar to Mandaeru in details. In the 'child library with a floored room', children can read books with bare fee and comfortable attitude, in the home-like atmosphere, laying on the floor.

The inner court of the library has a ground covered with stones, instead of lawn. The court is a stage for dance of imagination. The inner court exists for books. Wolcheongongdeok which was used to be conducted on the eve of Chinese lunar new year day according to the Korean traditions is to lay stepping stones. But, you should do it secretly. The stepping stones were very valuable essence in a community. The images of the stepping stones were used for architecture design. Water flows. And there are stepping stone across the stream.

The exterior detail are evocative of books

General view of the main entrance

A passage to the indoor and the courtyard

Courtyard

1 MULTIPURPOSE HALL
2 COURTYARD
3 READING ROOM
4 OFFICE
5 CIRCLE ROOM
6 DATA ROOM
7 PARKING LOT
8 CONTROL ROOM
9 STORE
10 CORRIDOR
11 BOOK CAFE
12 LECTURE ROOM

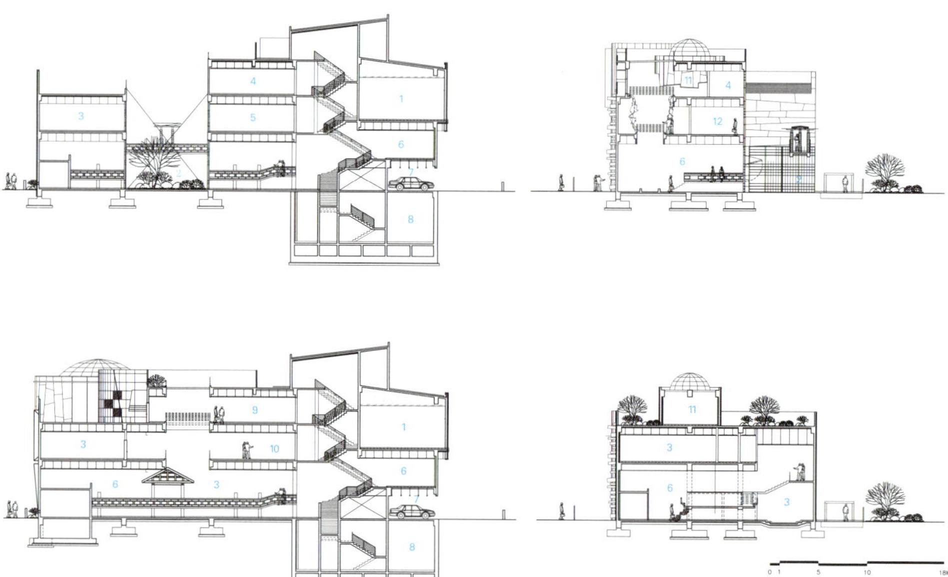

SECTION

Courtyard viewed from the leading room

General view of the leading room

Book cafe on the 3rd floor

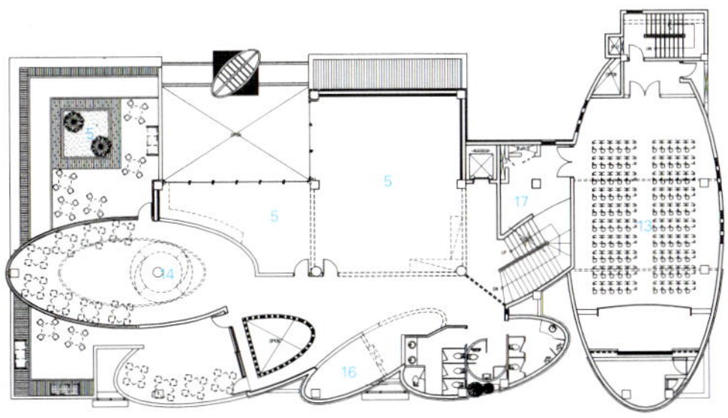

3RD FLOOR PLAN

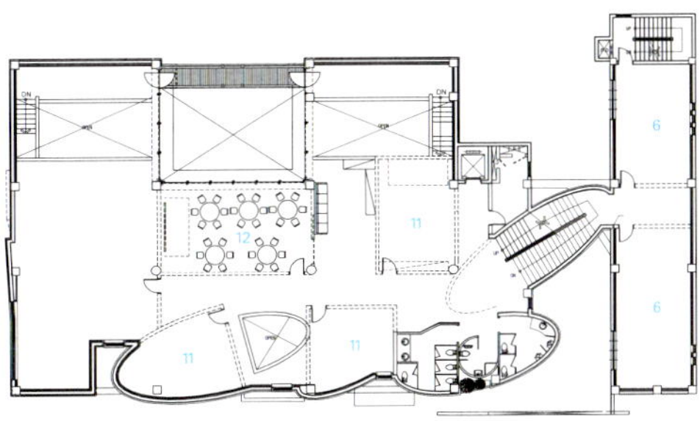

2ND FLOOR PLAN

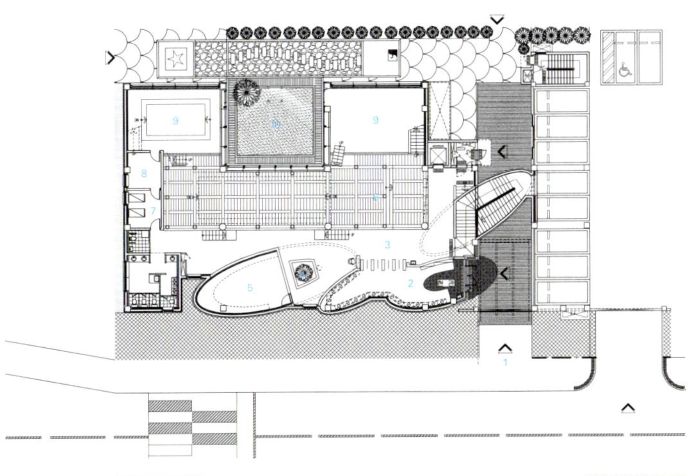

1ST FLOOR PLAN

1 MAIN ENTRANCE
2 ENTRANCE HALL
3 LOBBY
4 INFORMATION DESK
5 OFFICE
6 DATA ROOM
7 SLEEPING ROOM
8 ORALLY NARRATED FAIRY TALE ROOM
9 READING ROOM
10 COURTYARD
11 CIRCLE ROOM
12 LECTURE ROOM
13 MULTIPURPOSE HALL
14 BOOK CAFE
15 ROOF GARDEN
16 STORE
17 HALL

HAESONGWON_
DAEYANG SHIPPING TRAINING CENTER

Location Songjeong-ri, Mijo-myeon, Namhae-gun, Gyeongsangnam-do, Korea **Site Attribute** Control, Part Residential Development Promotion **Site Area** 5,768.00m² **Building Area** 1,039.19m² **Total Floor Area** 1,557.07m² **Building Coverage Ratio** 18.02% **Floor Area Ratio** 27.00% **Building Scope** 1~3F **Structure** RC **Exterior Finish** Titanium Zinc Sheet, Western Red Cedar Siding, Exposed Concrete, Native Rock **Interior Finish** Exposed Concrete, Marble, Fabric, Carpet Tile **Design Period** 2004.2~2004.10 **Construction Period** 2004.10~2005.12 **Architecture Design** Lee In-ho, Jung Min-hee, Kim Hyung-il | E.RAE Architects **Design Team** Jung Min-hee, Jeong Sin-ae, Kim Min-gi **Interior Design** E.RAE Architects, JOONGANG DESIGN **Client** Daeyang Shipping **Photographer** Kim Young-kwan(E.RAE Architects)

Overall view of the site with a river in front and a mountain behind

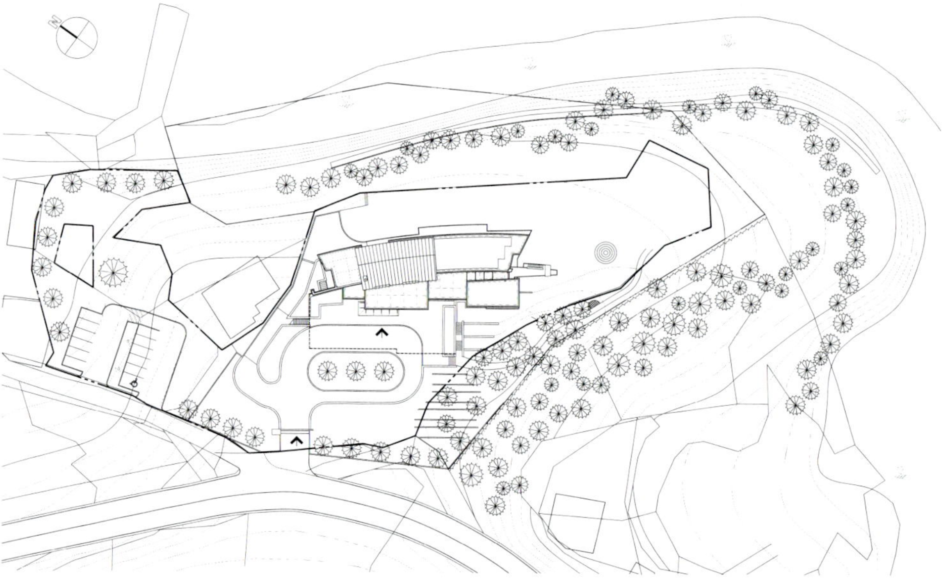

SITE PLAN

Beside the main entrance on the 2nd floor, there are other routes in the main facade which lead to the 1st floor of the rear side of the building.

General view of guest rooms, seen from the south-west.

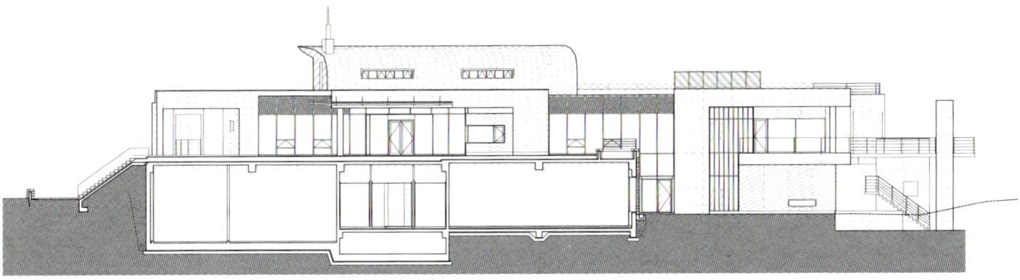

ELEVATION

Named after Hae-song, chinese characters meaning pine trees by a coast, Haesongwon is located on a grey rock bed surrounded by a blue coast in a forest where pine trees stand like a wall.

The site is sloped toward the sea and entrances are distributed on each floor using 5m level difference between the road and the sea levels of the site. The main entrance from the access road is on the second floor, while there is also an entrance to the coast on the first floor. To preserve oceanic scenery as much as possible, most programs are located on a single floor. Mass on the third floor is minimized in a double curved lines with walls and a roof in a uniform material. Programs requiring large area such as mechanical room and auditorium, etc. are located on the basement floor.

Curved lines in floor plans and irregular boxes create a twisted and sloped interior space. This drastic transformation of interior space constantly let us experience constantly changing quality of spatial magnitude according to daylighting condition and a view point, while textures and colors of materials give different feelinga depending on daylighting conditions.

After entering the main entrance on the second floor through a windbreak room, sloped against the building, we stand in the middle of stair hall with 3 stories open, feeling as if we were floating on a sea.

Open space of staircase become buffer space connecting the sky, the ground and the building. A small atrium facing a corridor in the south connects the first and the second floor in both visual and spatial ways, also inviting deep south daylighting. Window types and sizes are different in every location to accommodate the scenery of its neighborhood environment and the nature in various picture frames.

Instead of using standardized materials, materials requiring craftsmanship were used such as exposed concretes, dry stone walls, marble walls, red cedars and lightweight galvanized zinc plates. To emphasize sculptural image of the exterior stairway, exposed concretes were used. A vertical layer of columns and a diagonal layer of a staircase extruded from the wall make a harmonic image in contrast with the horizontal long building.

Exposed concretes and circular columns makes a harmony with pine trees erected upright from the ground. The scenery of its forest is overlapped with red cedar louvers through the window of a guest room.

The main facade in the northwest faces the southern sea, while the cube opens and closes repeatedly.

1 LOBBY
2 LAUNDRY ROOM
3 STORAGE
4 AUDITORIUM
5 DORMITORY
6 CAFETERIA

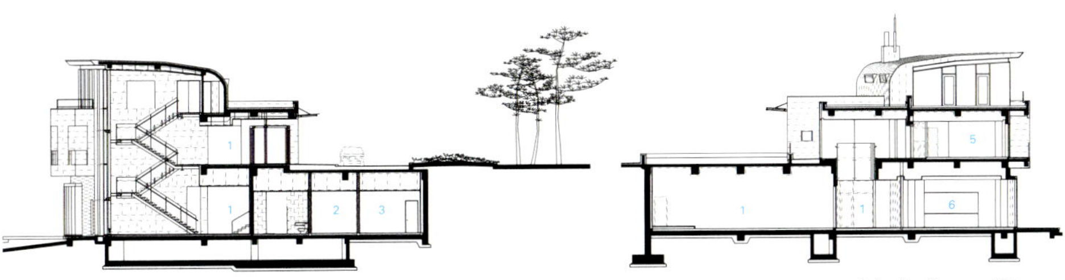

SECTION

Exterior staircase makes us feel closer to its neighborhood environment, the southern sea, while exposed concrete emphasizes its sculptural image.

Lobby on the first floor in the southeast. Curved line and straight line are in contrast, while repetitive sequence of space increases dynamic atmosphere of the spaces.

Glass staircase open throughout three floors, starting from the first floor.

Dining room on the first floor, which makes a panoramic view of pine trees though glass windows.

Vip meeting room on the 2nd floor

Exterior stone materials are exposed through windows in the auditorium on the first floor.

North-west view of the building from an outdoor deck on the 3rd floor

Interior of recreation room on the 3rd floor, where a lot of glasses were used to invite natural scenic view.

1 LOBBY
2 AUDITORIUM
3 LAUNDRY ROOM
4 STORAGE
5 AUDIO ROOM
6 MACHINE ROOM
7 ELECTRIC ROOM
8 KITCHEN
9 CAFETERIA
10 SEMINAR ROOM
11 GUEST ROOM
12 OFFICE
13 RECREATION ROOM
14 OUTDOOR DECK

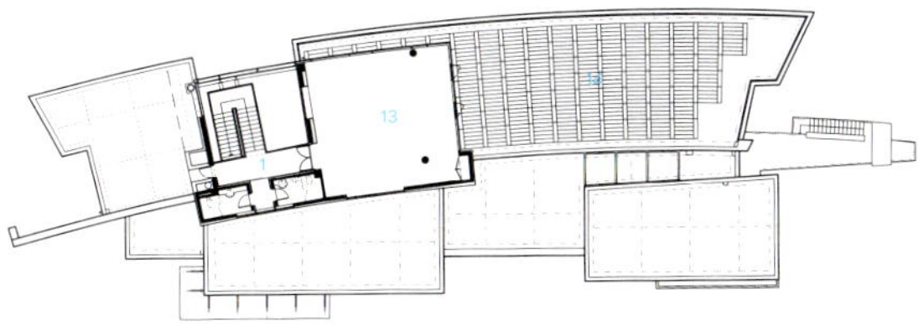

3RD FLOOR PLAN

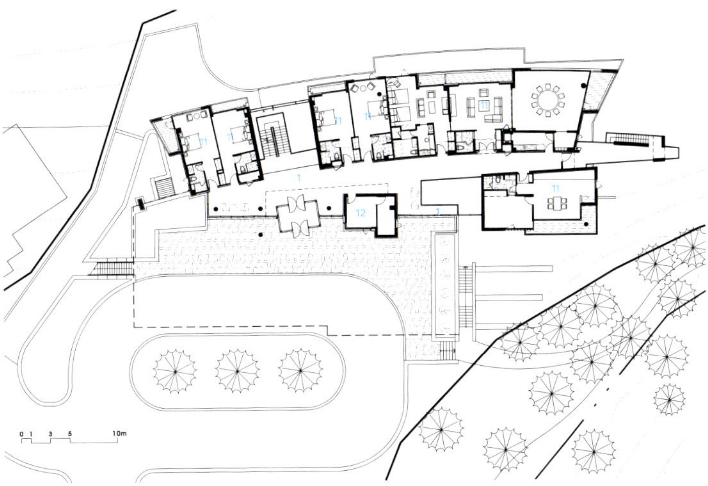

2ND FLOOR PLAN

1ST FLOOR PLAN

*EDUCATION COMPETITION

NATIONAL DAEGU SCIENCE MUSEUM

Location Sang-ri, Yuga-myeon, Dalseong-gun, Daegu, Korea **Site Area** 117,355m² **Building Area** 16,133m² **Total Floor Area** 23,464.95m² (Ground 19,515.02m², Underground 3,949.93m²) **Building Coverage Ratio** 13.75% **Floor Area Ratio** 16.63% **Building Scope** Science Center_B1~4F, Cosmos Eye(Dormitory)_5F **Maximum Height** 32.6m **Structure** Steel Truss, Steel, SRC **Exterior Finish** AL Panel, Low-E Pair Glass **Interior Finish** Artificial Marble, Carpet Tile, Eco Water-Based Paint **Architecture Design** Jeong Young-kyoon | HEERIM Architects & Planners **Design Team** Bae Seong-rok, Ko Mak-seon, Choi Hyeon, Kim Dong-uk, William Craig, Lee Mi-jin, Lee Hye-jin, Song Sang-hwan, Kim Hyo-jin, Kim Ho-gyeom

DESIGN CONCEPT

· Eco + Digital = Eco Digital Science Park
  - Nature(Perfection of Network): By making a science park connected with the central park, the greens to exist scattered within the technopolis are contained in the green network associated to 'DGIST(Daegu Gyeongbuk Institute of Science & Technology) - central park - science park - Bisel Mountain' and the green network of the whole technopolis shall be finished.
  - Working Nature/ Science: The outer space is a nature for the rest and does not exist in itself and becomes a working nature . The water purification process of the natural purifying plants in swamp is a place for a vivid ecology experience and an Algae, a kind of green algae, generates oxygen and hydrogen by the chemical action and they are used as an energy source in the land.

· The nature of the park is a nature to function like so, and becomes a science itself.
  - Science Museum/Science Park: Getting out of the existing exhibition method to make a visual information transfer, it extends its sphere to the outer space. In the swamp, RG farm, eco digital band, living machine, a scientific concept is realized and it is a place for a vivid scientific experience and a sphere of the scientific building is expanded not to a single building but whole science park.

CONCEPT OF SHAPE: DIGITAL ECOLOGY Substructure mass landscaped to adapt to the existing topology and the urban flow symbolizes a nature and the flying upper stories open to the central point of the city and the central park symbolizes a future-oriented scientific technology.

ARCHITECTURAL SPACE CONSTITUTION (CONTINUOUS LOOP): FLOW OF CONTINUOUS WATCHING TRAFFIC LINE

- Centered at the central hall which 4 stories are open, it is designed for a skip floor type.
- Through a moving walk, keeping a flow of continuous watching traffic line
- By setting up a nude elevator, observing a large-scale central hall and an inner space

PLANNING

- Dividing a sphere for research and office work and a sphere for exhibition and keeping an independence
- By arranging public beneficiary facilities(restaurant, museum shop, lounge and resting deck) between exhibition facilities, maximizing an efficiency for function to watch an exhibition.
- By planning as a large scale open exhibition space, possible to design super-variable exhibition

ELEVATION PLAN

· Nature + Science
  - Expression of the elevation to substitute a digital to mean a nature for lower part and a science for upper part
  - Natural Facade: It is planned for a sympathy with the nature through a transparent glass and a fusion with the nature through greening the roof in the lower part.
  - Gradient Facade: Presentation for the elevation possible for continuous change by making a shape for a scientific digital image through a geometry in the nature and changing a transmittance of a light with a solid and a void of the modularized unit.

1.건축설계 개요 및 개념
배치 프로세스
인공 vs 자연
인간 vs 기술
인공 vs 자연/ 인간 vs 기술
생태 에너지 생산 시스템 구축
진입공간 및 과학관 배치
건물개요
대지면적
건축면적
32.6M
13.75%
기타시설물 개요
주요부분 마감
설비개요
주차개요
조경개요
기타사항

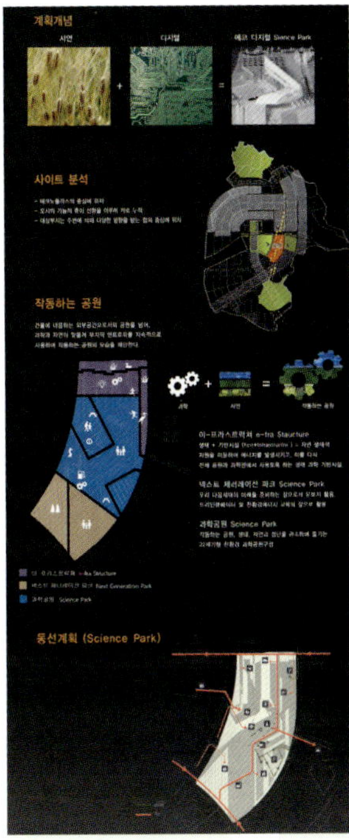

3.배치 및 옥외공간계획

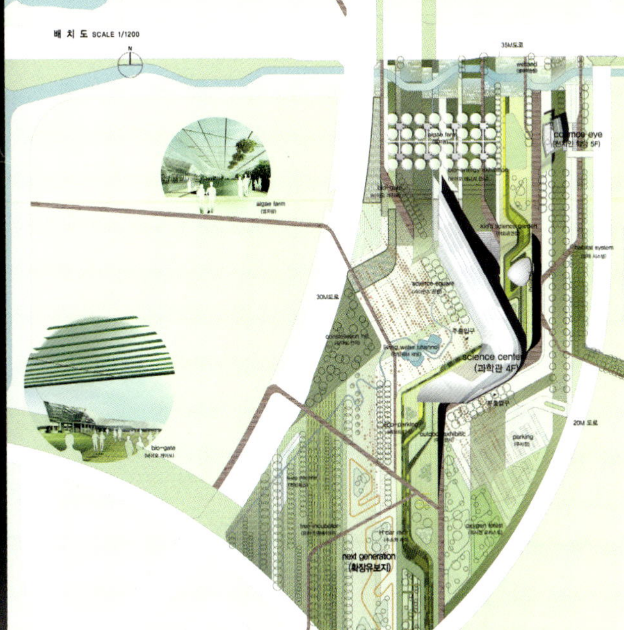

4.각층 평면도

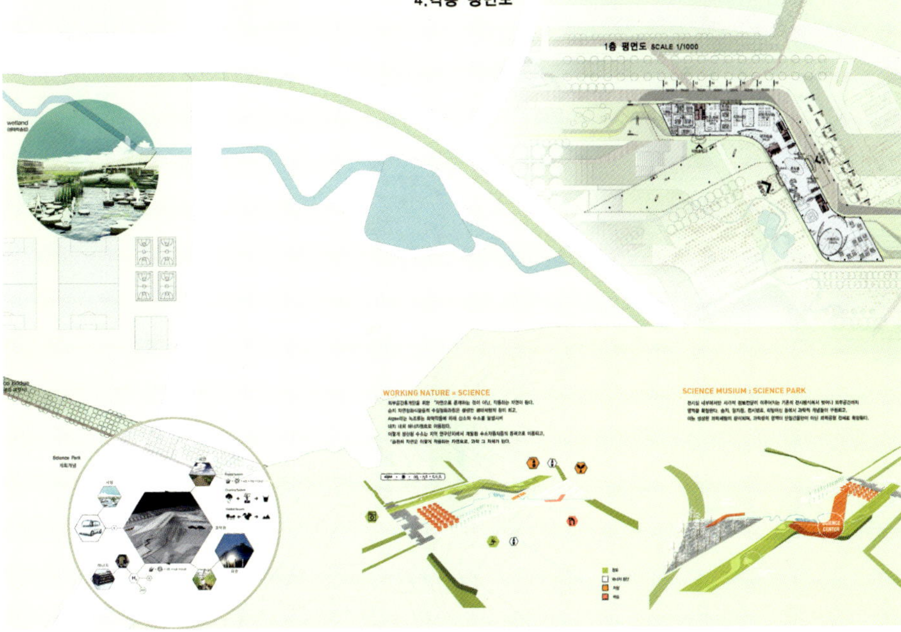

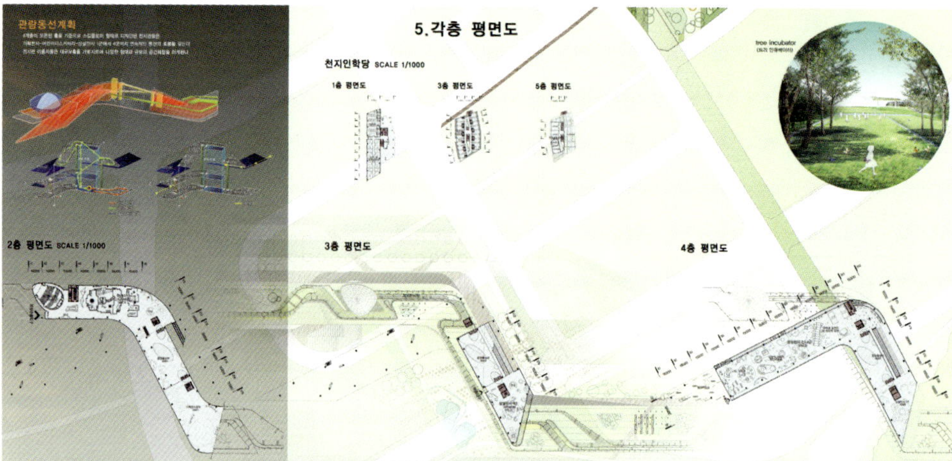

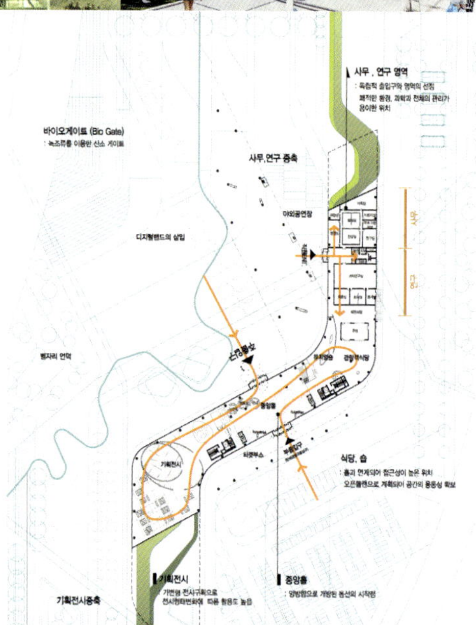

- 긴 선형매스의 지루함을 감소시키기 위해 중앙홀을 기준으로 사무,연구,휴게 등의 시설들을 조닝하고 반대편에 기획전시를 배치하여 관리, 일반동선 및 전시동선을 분리
- 양방향으로 오픈된 대규모 중앙 홀 계획으로 대지흐름, 진입의 연속성 유지
- 계획 초기단계부터 증축을 고려한 배치

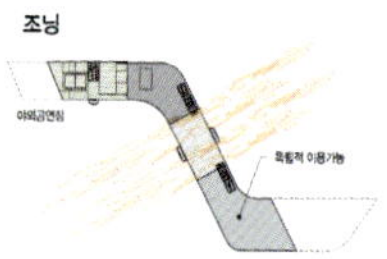

1ST FLOOR PLANNING

- 동시 이용자수가 많은 4D디스커버리는 독립적 진입이 가능하도록 2층에 배치
- 활동량이 많고 관리가 필요한 어린이영역은 1층 관리동선과 연접배치하여 관리편의 도모
- 충분한 층고의 확보와 오픈플랜의 도입으로 전시형태의 변화 등에 능동적 대응
- 충분한 증축부지 확보

조닝

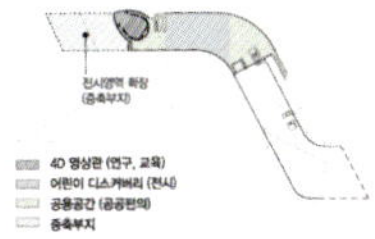

동선

*동시 이용자 수가 많은 4D영상관 독립배치
*중앙홀과 연접배치하여 어린이 영역 관리 용이

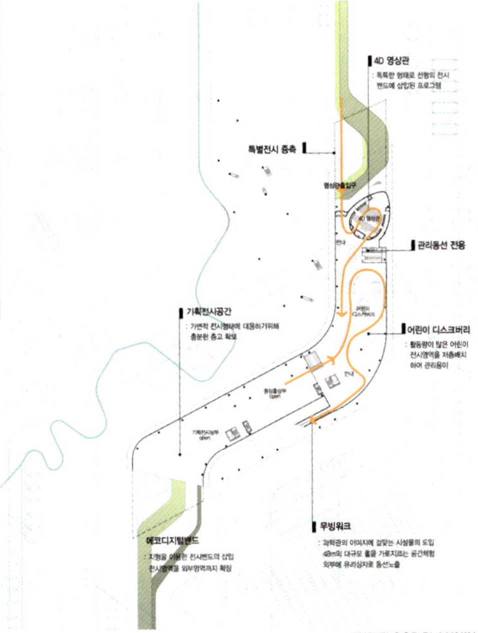

2ND FLOOR PLANNING

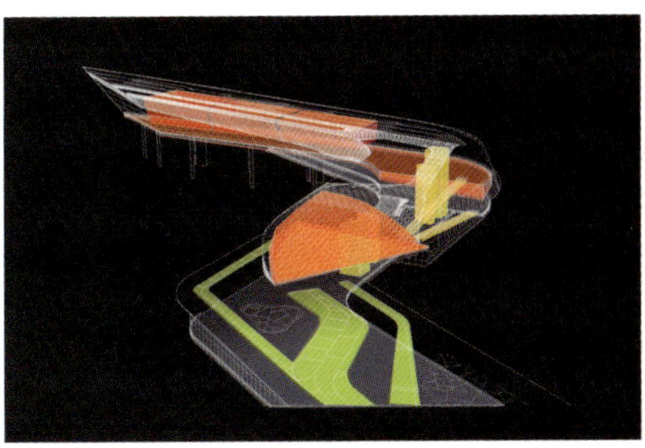

- 스킵플로어로 구성된 상설전시공간으로 중앙홀의 무빙워크를 통해 대공간 체험
- 오픈플랜 구성으로 공간의 융통성을 극대화하여 전시 변화에 능동적 대응
- 상설전시 1군에서 에코 디지털밴드로 바로 연결되어 외부체험 전시공간까지 전시공간 확장
- 계획 초기단계부터 증축을 고려한 배치

전시영역의 확장

*디지털에코밴드 : 전시밴드의 구성으로 전시영역을 옥외로 확장

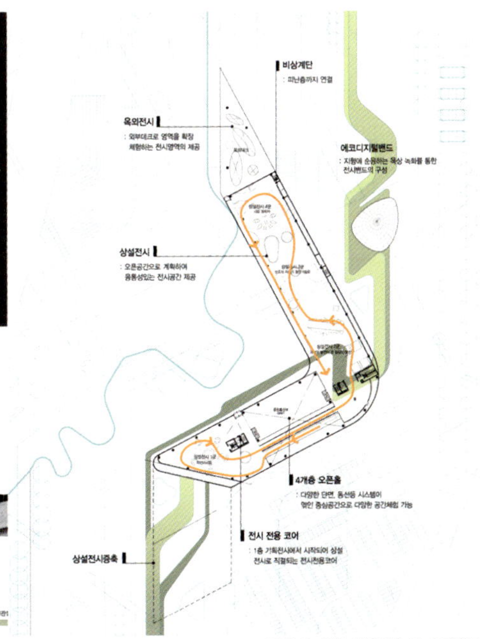

3RD & 4TH FLOOR PLANNING

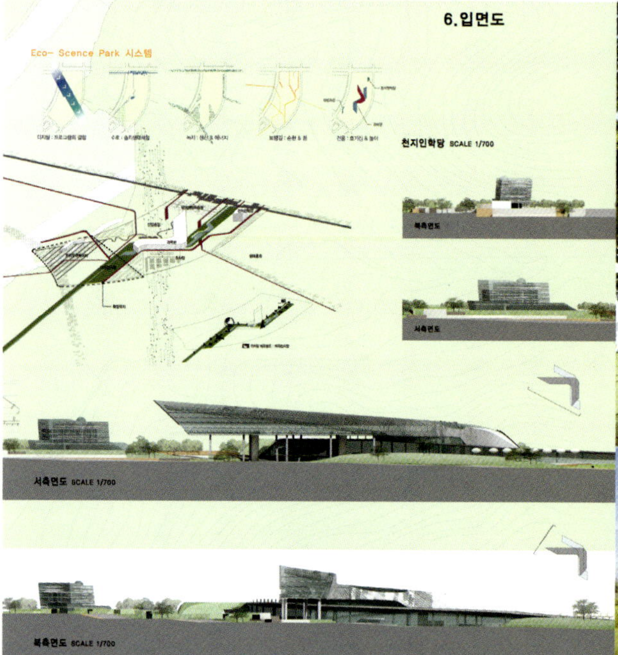
6.입면도
Eco- Science Park 시스템
천지인학당 SCALE 1/700
북측면도
서측면도
서측면도 SCALE 1/700
북측면도 SCALE 1/700

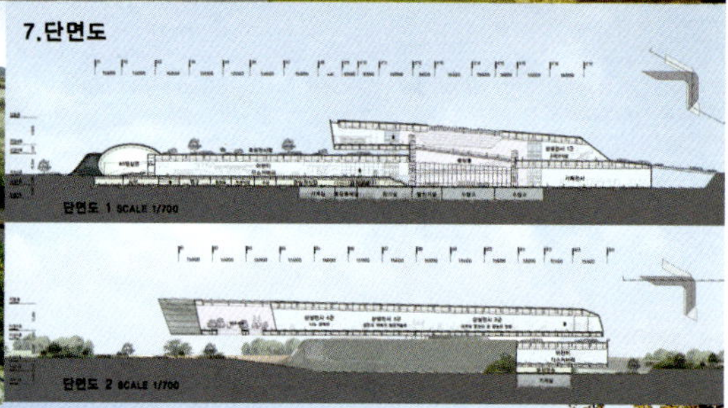
7.단면도
단면도 1 SCALE 1/700
단면도 2 SCALE 1/700

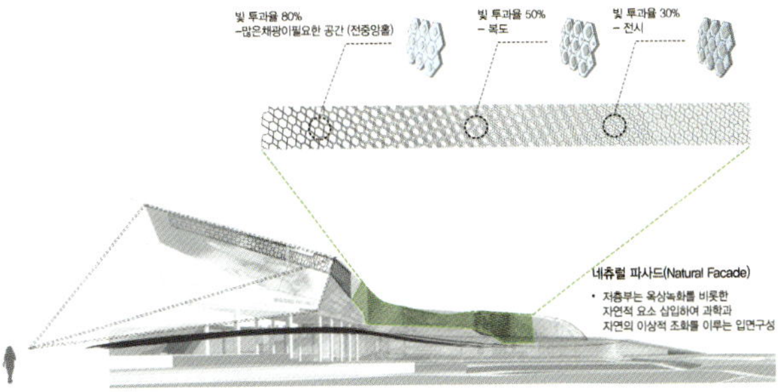

그레디언트 파사드(Gradient Facade)

- 모듈화된 유닛의 솔리드와 보이드를 변화시켜 공간의 기능에 적합한 채광을 확보
- 전체적인 입면의 이미지 역시 솔리드에서 보이드로 자연스럽게 변화해가는 과정을 표현
- 모듈화된 기하학을 통해 과학적 디지털 이미지 형상화

네츄럴 파사드(Natural Facade)

- 저층부는 옥상녹화를 비롯한 자연적 요소 삽입하여 과학과 자연의 이상적 조화를 이루는 입면구성

ELEVATION PLANNING

계획의 주안점

- 중앙홀을 중심으로 전시공간들이 양방향으로 배분되어 무빙워크를 통해 연속적 전시관람 가능
- 1층에 휴게, 연구, 사무공간등 층고가 낮은 시설을 조닝하고, 상부 어린이 디스커버리 4D영상관등의 접근성을 높임
- 긴 평면구성을 보완하기 위한 코어의 적정배분으로 접근성 높임

수장시설

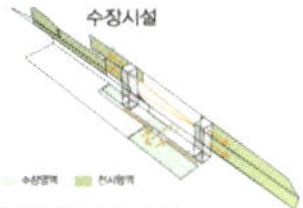

* 명확한 동선계획으로 물품, 반출입에 있어서의 기능성과 안전성을 확보

개방

* 상부 오픈된 개방형 로비 구성으로 풍부한 공간과 인지성 부여

부유

* 부유하는 매스는 과학관의 미래 지향적 이미지를 표현

영역

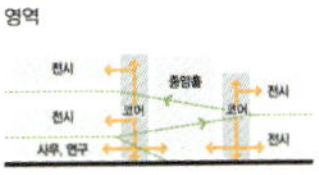

* 중앙홀을 전시공간의 도입부로 활용하여 상층부로 자연스런 진입유도

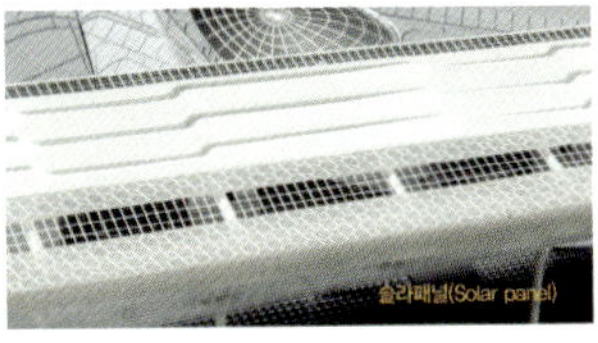

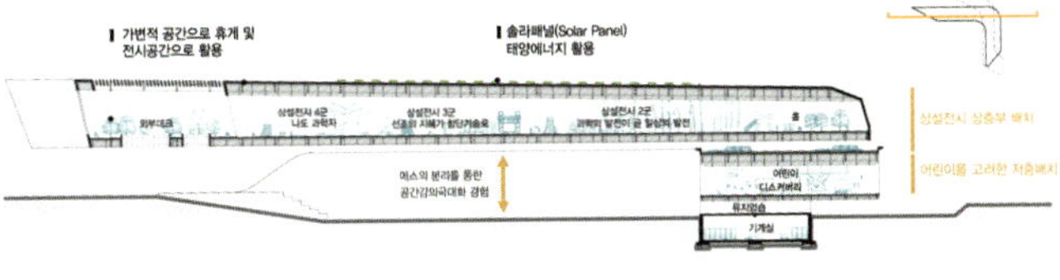

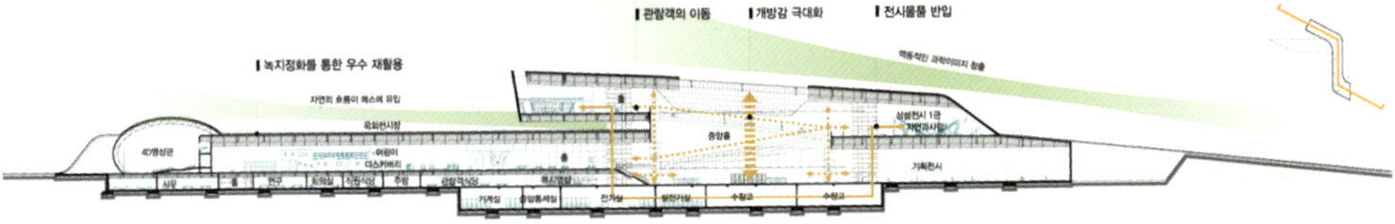

SECTION PLANNING

>> EDUCATION

ULSAN HUMAN RESOURCE SUPPORTING CENTER

Location Ok-dong, Nam-gu, Ulsan, Korea Site Attribute Natural Green Site Area 40,656m² Building Area 7,019.56m² Total Floor Area 23,032.06m² Building Coverage Ratio 17.27% Floor Area Ratio 45.84% Building Scope B1~4F Structure RC, Steel Parking 197 Cars Exterior Finish Pair Glass, Titanium Zinc Plate, Oxidized Copper Plate Architecture Design Kim Chung-young, Koo Ja-sun | HEEHYUN Architects + Kim Won-hyo | WONRYEON Architects Design Team HEEHYUN_Gu Tae-hyeong, Yang Mi-sun, Choi Hyeong-yeong, Choi Eun-sil, Kim Seong-hoe, Han Dong-hui, Lee Jeong-eun / WONRYEON_Hong Seung-tae, Oh Yong-du

ECO-VALLEY(NATURE+MAN) & SCIENCE-VALLEY(NATURE+MAN)

FORM OF READING THE SITE Making the most use of the existing topography and the natural environment, and even putting an emphasis on the method to utilize them, the building is naturally put in the valley between the ridge connected with the district to preserve trees and the Samho mountain. The natural factors(greens, wind, water), the building, and the man make a natural combination by communicating with the nature installing a connecting road with the original nature, making a way for wind between the creation learning center and the future human resource center, accepting the nature, forming the water naturally by blocking a low swampy place and using them as the learning center and the reusing of supernatant liquid.

RECOGNITION OF FORM In the roadside to link Ulsan and Seoul, an oval type is made in order to emphasize a constant frontness and a recognition is emphasized.

RECOGNITION OF PLAIN The threads by roof-type plain constitution enclosed with the central garden and the atrium is closely connected. The creation learning center is made to bring up the scientific dream through research and experiment through an inducement of curiosity by piling up the layers as like peeling an onion's skin by piling up a repetitive orbit of the solar system and the future human resource center is an open plain shape and it keeps a continuity of an information center(tentative name) scheduled for extension in the future and emphasizes a dynamics to spread out to the future. The connecting deck connects the creation learning center and the future human resource center harmoniously, emphasizes an independent function of the plannerterium, accepts the visual axis and puts an emphasis on the connectivity considering a future extension.

RECOGNITION OF ELEVATION With a theme of the Eco-Shadow, an elevating expression and an eco-friendly recycling energy to closely connect nature, man and science are actively introduced as an elevated factors and used as an educational purpose. A close connection introduces various elevating factors of a vertical and horizontal sphere and a progressive dynamics is expressed through a slant factor(an ecological way). Through a shade, a memory into an eco-friendly ecological space is recreated on the elevation and for the part of eco-friendly recycling energy, 4 factors are reflected into an elevation and they are used as an educational purpose.

- Photovoltaic Power Generation System: installing solar panel on the elevation, generating an electricity and using for heating
- Photo-duct system to use a light: A natural light is accepted with a prism, transmitted into a toilet or an underground by the photo-duct and plays a role for a sterilization through an ultraviolet as well as a light.
- System to use reusing liquid and rain water: the water and the rain water of an artificial low swampy place(ecological learning place) are collected and used as water for rinsing and water for cleaning.
- Heating & cooling system to use geothermy: by means of a difference of the ground temperature, it is used as a cooling in the summer and a heating in the winter.

배 치 도
축척 : 1:600
부출입구
주출입구
3m도로확보
기상관측원(53.0)
바람의 언덕(53.0)
암석원(52.0)
수목보존지역
아귀마당(43.0)
(상징조형물)
생태연결통로
(eco-bridge)
바람의 계곡
(wind-valley)
창의학습관(4F)
미래인재관(4F)
(가칭)정보관 증축예정지
생태길
(eco-road)
물의 언덕
생태체험학습장
(제안사항)
버스하차장
설계개요
사 업 명 울산인재육성지원센터 신축 설계용역
대지위치 울산광역시 남구 옥동 894번지 일원
지역지구 자연녹지지역
용 도 교육연구시설
대지면적 40,656.0 m²
건축면적 7019.56 m²
연 면 적 23,032.06 m²
(지하 : 4,395.38 m², 지상 : 18,636.68 m²)
건 폐 율 17.27 %
용 적 율 45.84 %
층 수 지하1층, 지상4층
구 조 철근콘크리트조,철골조
조경면적 12,237.46 m² (30.01%)
주차개요 197대 (장애인 17대)
외부마감 복층유리 ,티타늄아연판, 녹청산화동판
동선계획
외부공간계획
생태연결통로(Eco-bridge)
어귀마당
바람의 계곡(WIND-VALLEY)
생태길(Eco-road)
물의 언덕
에코 밸리(Eco-valley)
사이언스 밸리(Science-valley)
형태의 인지성
자연과의 어우러짐
삼호산
증축의 고려

생태벨트(Eco-Belt)의 형성
지하1층평면도
4층평면도
3층평면도
평면계획
1층평면도
2층평면도

정면도
배면도
좌측면도
우측면도
입면계획
단면계획
횡단면도
종단면도

ULSAN HUMAN RESOURCE SUPPORTING CENTER

Location Ok-dong, Nam-gu, Ulsan, Korea **Site Attribute** Natural Green **Site Area** 40,656m² **Building Area** 6,377.46m² **Total Floor Area** 22,016.46m² **Building Coverage Ratio** 15.69% **Floor Area Ratio** 41.43% **Maximum Height** 26.00m **Building Scope** B1~4F **Structure** SRC, RC **Parking** 224 Cars **Exterior Finish** Meta Panel, T24 Low-E Color Pair Glass, Western Red Cedar(Preservative Treatment) **Architecture Design** Heo Dong-yun | SANGJI Environment & Architects + Kim Jin-han | GROUP CS Total Architects & Consultants + Kim Ki-soo | Donga University **Design Team** SANGJI_Go Seong-ryong, Jeong Jae-yeong, Im Ho-taek, Park Sun-yeong, Jeong Hae-deok, Gwak Ji-hye / GROUP CS_Go Yeong-ju, Lee Tae-bong, Ryu Ju-yeong

ANALYSIS OF SITE While the object site is located at the node of the Munsu road and the southern beltway in Ulsan and has a strong positional recognition and a character of place, it holds a weakness which a recognition of the site, an existence of the place and a horizontal elevation of the building showed at the Human - View become extinct because it is lower than the level of the nearby road. In order to understand such a condition of the site, the understanding on the site is begun through an investigation to the recognition of the building, two dimensional (human, social, geological) characteristics which the site owns and three dimensional analysis which landform and shape hold.

DIRECTION OF PLAN The direction of plan aims to classify a program which the human raise-up support center owns classified as three great causes and to accept a wide scientific knowledge, raise a creative human resource and transmit new scientific information. Also, it aims to make a complex culture space (Public Square) to achieve a synergy effect with the local vision and culture.

BLOCK PLANNING The main arrangement plan has an emphasis to contain a natural line analogized from the flow of the original topography and make a method to show a recognition of the building an architecture. The volume of the building's mass and the two dimensional(horizontal) form are selected by a shape of a valley which a site owns and a natural force which the valley holds, and the three dimensional (vertical) design is constituted as a method to show the original topography buried and a method to draw the valley into the building.

ELEVATION PLAN A vertical uptrend, a vision toward the future and a recognition of the building are elevated and expressed and it is planned for the Skin of the elevation with numerical stars' shapes in the night sky a basic Motive.

PLAN FOR OUTDOOR SPACE It reanalyzes the original nature's line showed from the site as an image of the circuit of the informational and scientific factors and gives an outdoor space a characteristic and a property with topography and overlapped lines and divided.

I wish that learning of knowledge, mutual communication of information and eco-friendly design as a public space, all these are realized in the valley to bring up the human resources.

■설계개요
구 분 | 설계내용
건설개요 명 칭 울산인재육성지원센터 신축 설계경기
대지 위치 울산광역시 남구 옥동 894 번지 일원 (울산제일고등학교 인근)
지역 지구 자연녹지지역
대지 면적 40,656.00 m²
건축 면적 6,377.46 m²
연 면 적 22,016.46 m²
지상층연면적 16,795.54 m²
구 조 철골철근콘크리트조, 철골조 / 철근콘크리트조
주 용 도 교육연구시설
건 폐 율 15.69 % (법적기준 : 20% 이하)
용 적 률 41.31 % (법적기준 : 100% 이하)
규 모 지하1층, 지상4층
최고 높이 26.00m
주차개요 법 정 93대
계 획 자주식주차 총 224대 ([illegible])
■조경 계획
■외부공간계획
■동선 계획
보행자 안전 및 쾌적성을 위한 보차분리
내외부를 소통하는 보행자동선 구축
다양한 외부공간을 체험할 수 있는 순환동선 계획
대지전체를 순환하는 차량동선 계획
미래인재관
창의과학관
■정면성
■친환경적 배치
■확장성

234

Incubator In Valley 울산 인재 육성지원센터
■ 계획개념
■ 부지현황
■ 대지분석
Incubator Of Ulsan
Knowledge Bank
Public Square
Science Valley
■ 지하1층 평면도
■ 지상1층 평면도
■ 과학관 지상2층 평면도
■ 인재관 지상2층 평면도
■ 과학관 지상3층 평면도
■ 인재관 지상3층 평면도
■ 과학관 지상4층 평면도
■ 인재관 지상4층 평면도
■ 입면계획 방향
■ 북측입면도
■ 동측입면도
■ 대지횡단면도
■ 대지종단면도
■ 친환경 계획 방향
■ 태양에너지
■ Double Skin
■ 우수활용
■ Eco – Void
■ 생태 체험장 (아름드리 학습장)
■ 건축물 내부 생태경관
(과학관 2층 4층, 인재관 1층 옥상층)
■ 단면계획 방향

GWANYANG LIBRARY & WELFARE COMPLEX

Location Gwanyang 1-dong, Dongan-gu, Anyang-si, Gyeonggi-do, Korea **Site Attribute** General Resident **Site Area** 4,777.10m² **Building Area** 2,484.16m² **Total Floor Area** 15,348.33m² **Landscape Area** 858.40m² **Building Coverage Ratio** 52.07% **Floor Area Ratio** 167.37% **Building Scope** B2–4F **Maximum Height** 21.30m **Structure** RC **Parking** 193 Cars **Exterior Finish** Extruded Cement Panel, Color Pair Glass, Aluminium Composite Panel, Wood Louver **Interior Finish** Stone, Fabric Panel, Gypsum Board, Paint, Interior Finish **Architecture Design** Cheon Joon-ho, Hwang Kyung-taek | H.L Engineering **Design Team** Yun Taek-yong, Lee Sang-jin, Song Ji-kyeong, Choi Yeong-jin, Park Jun-hyeon, Kim Jeong-hwan, Jeong Seung-ho, Choi Myeong-hwan

Anyang, the city of lifelong education where work, leisure and study are harmonized. In this project, the designer proposes Gwanyang library and welfare complex a new concept one calls another to create a space and harmonized with nature hold the identity of 21st Anyang. Through the emphasized natural flow of people, the previously vacant site is occupied by spaces that provide different function to make a place of new experience and learning, which preserves the memory of the site and propose a community complex.

BASIC PLAN DIRECTION

· Community
  - Inducing the interaction and communication among residents by supplementing new community facilities in Gwanyang 1-dong such as a library and a welfare complex.
  - Inducing the participation of residents through programs connected to the school and surrounding residential groups.

· Natural
  - Establishing two open areas that reflecting the identity of the existing open space on the site.
  - Introducing an environment-friendly plan and an atrium inside the mass by expanding the two open spaces.

· Sympathy
  - Conducting the role of the center through Gwanyang library and the welfare complex, which are the sole complex programs in Anyang.
  - Creating a landmark and establishing an efficient circulation flow through the programs and layout planned in consideration of surrounding contexts.

SPECIALIZED SUBJECTS

- Offsetting and eliminating the source of noise in the background of the library using three devices for a quiet atmosphere of a reading room.
- Introducing a multipurpose room and variable walls for the efficiency of spatial use.
- Connection according to the relation between surrounding contexts and inner programs.
- Encouraging the young adults to use of the reading room by installing a bridge.
- Proposing a new relaxing area by preserving the existing gate ball ground and reinforcing the function of relaxation.
- Ensuring the identity by maintaining the function of relaxation of the rooftop garden of the library.

LAYOUT PLAN We focused on maximizing an open area by ensuring an open view from the road side through a relatively efficient daylight program and by maintaining and providing the existing open space. We also emphasized the harmonization with the surroundings and the sense of frontage by adapting the layout to the existing urban axis and rational design that opens up the entrance.

FLOOR PLAN

Efficient Zoning + Eco System + Digital Community

- Effective zoning of the floor that separates the areas for the users and staff.
- Harmony of the light, wind and the space.
- A new place of communication of analogue and digital.

관양동 주민에게 아낌없이 주는 복합커뮤니티 공간
Giving tree
소음
향, 조망
도로체계 및 주변상황
종합분석
건축개요
사 업 명 : 관양도서관 및 복합복지센터 건립공사
대지위치 : 안양시 동안구 관양1동 1407-1번지
지역지구 : 제1종 일반주거지역
대지면적 : 4,771.10㎡
건축면적 : 2,484.16㎡
용 적 률 : 167.38%
구 조 : 철근 콘크리트 구조
층 수 : 지하2층, 지상4층
최고높이 : 21.3m
주차대수 : 193대

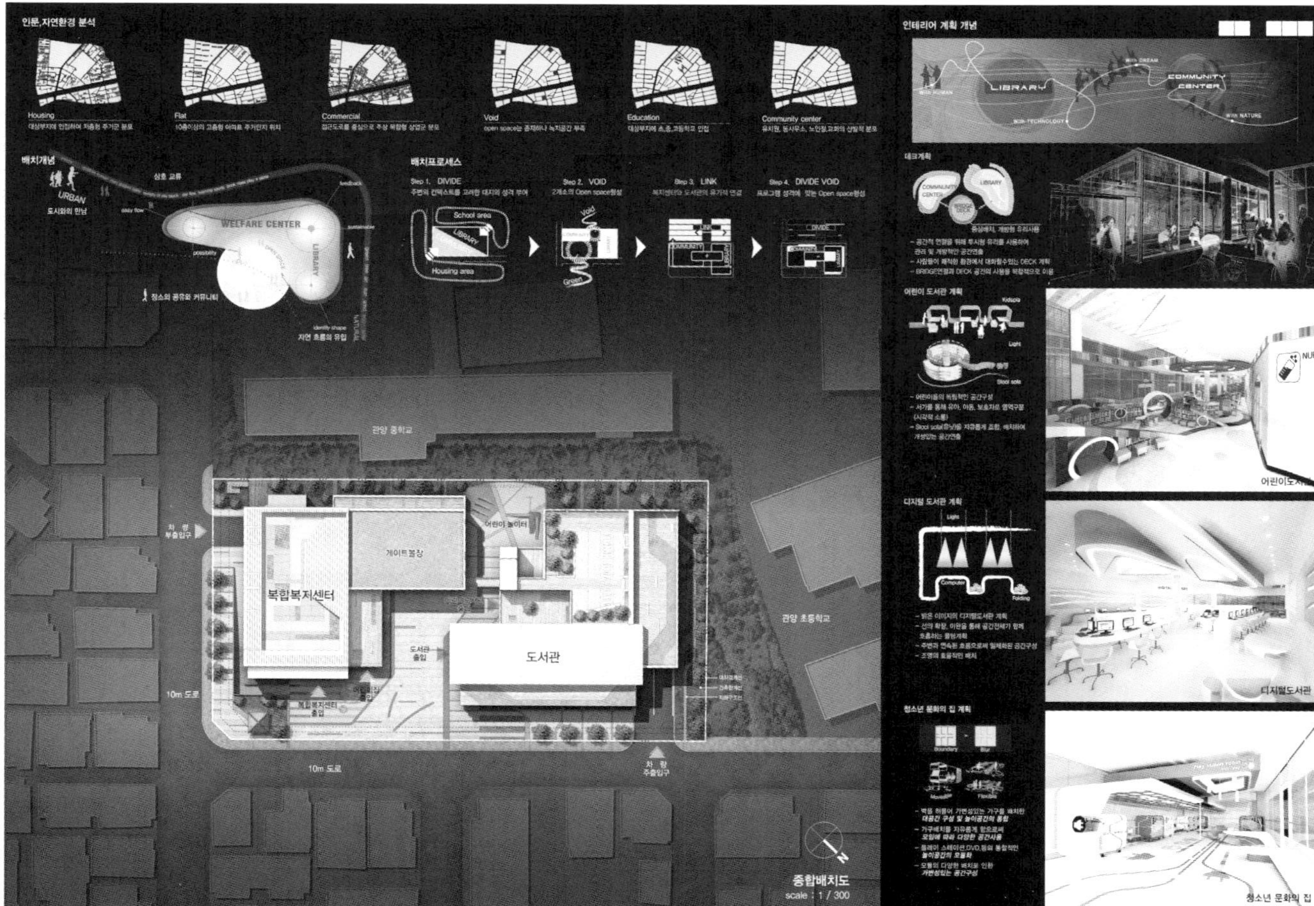

인문,자연환경 분석
Housing
Flat
Commercial
Void
Education
Community center
배치개념
URBAN
WELFARE CENTER
LIBRARY
배치프로세스
Step 1. DIVIDE
Step 2. VOID
Step 3. LINK
Step 4. DIVIDE VOID
School area
Housing area
관양 중학교
복합복지센터
게이트볼장
어린이 놀이터
도서관
관양 초등학교
10m 도로
종합배치도
scale : 1 / 300
인테리어 계획 개념
LIBRARY
COMMUNITY CENTER
데크계획
어린이 도서관 계획
디지털 도서관 계획
청소년 문화의 집 계획
어린이도서관
디지털도서관
청소년 문화의 집

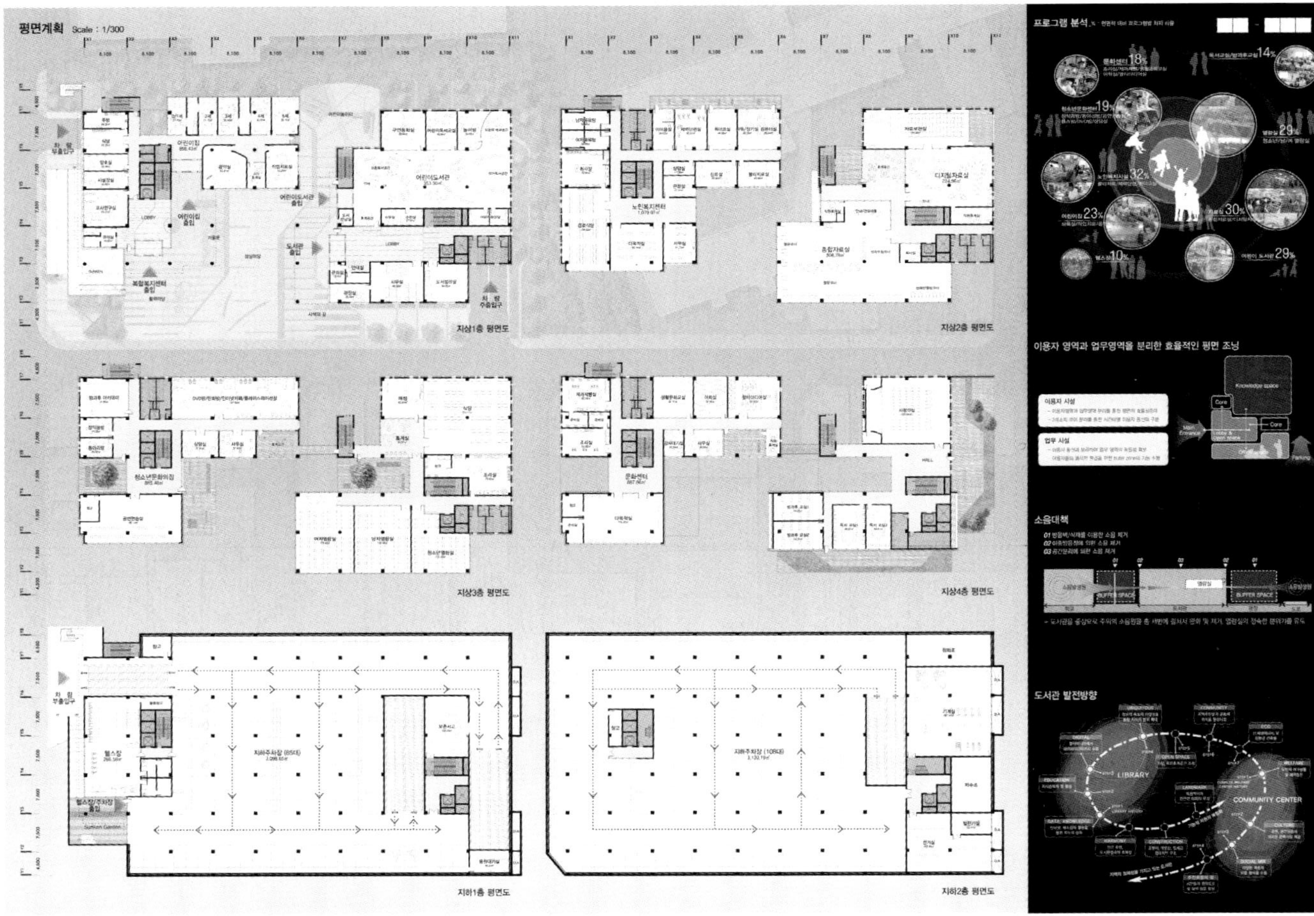
평면계획 Scale : 1/300
지상1층 평면도
지상2층 평면도
지상3층 평면도
지상4층 평면도
지하1층 평면도
지하2층 평면도
프로그램 분석
이용자 영역과 업무영역을 분리한 효율적인 평면 조닝
이용자 시설
업무 시설
소음대책
01 방음벽/식재를 이용한 소음 제거
02 이중벽등에 의한 소음 제거
03 공간분리에 의한 소음 제거
BUFFER SPACE
도서관 발전방향
LIBRARY
COMMUNITY CENTER

매스 프로세스
SCHOOL
HOUSING
ECO
Step 01 /
SOLID
Step 02 /
DIVIDE
Step 03 /
CURVED SPACE
Step 04 /
VOID
입면계획 주안점
주안점
Commercial
Residence
School Zone
SITE
Residence
Nature
Process
네트워킹
정보의 교류를 통한 새로운 지식의 양산
INFORMATION
LIBRARY
자연의 인입
공간구성도
Pattern Design (Digital + Analog = Digilog)
Digital
Nature
Analog
활력마당 진입부에서 본 전경
입면계획 scale : 1/300
정면도
좌측면도
배면도
우측면도

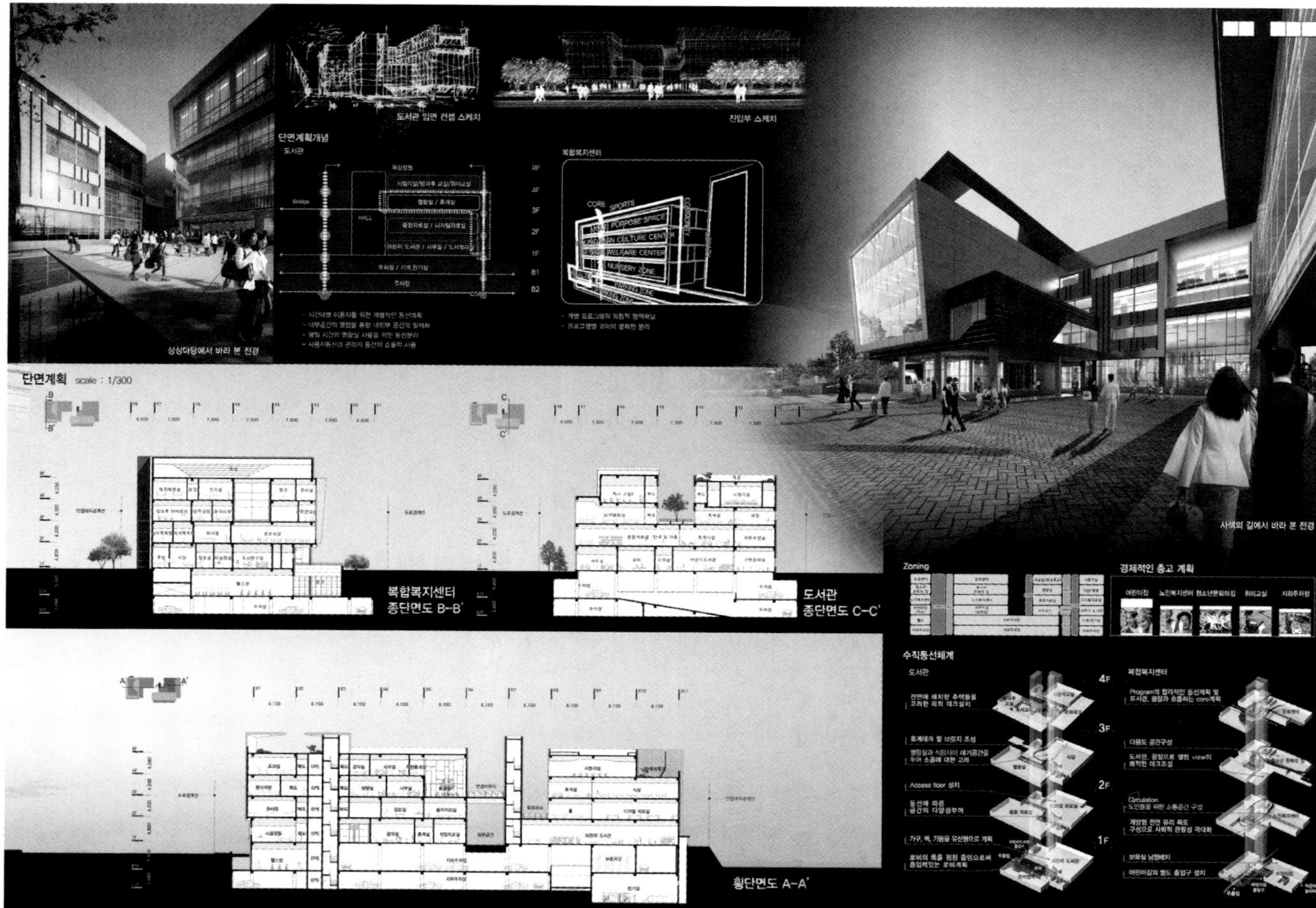
도서관 입면 컨셉 스케치
진입부 스케치
단면계획개념
도서관
복합복지센터
CORE SPORTS
MULTI PURPOSE SPACE
URBAN CULTURE CENTER
WELFARE CENTER
NURSERY ZONE
PARKING ZONE
상상마당에서 바라 본 전경
단면계획 scale : 1/300
복합복지센터
종단면도 B-B'
도서관
종단면도 C-C'
횡단면도 A-A'
사색의 길에서 바라 본 전경
Zoning
경제적인 층고 계획
어린이집
노인복지센터
청소년문화의집
취미교실
지하주차장
수직동선체계
도서관
복합복지센터
4F
3F
2F
1F
Access floor 설치
Circulation

동선 및 주차계획도

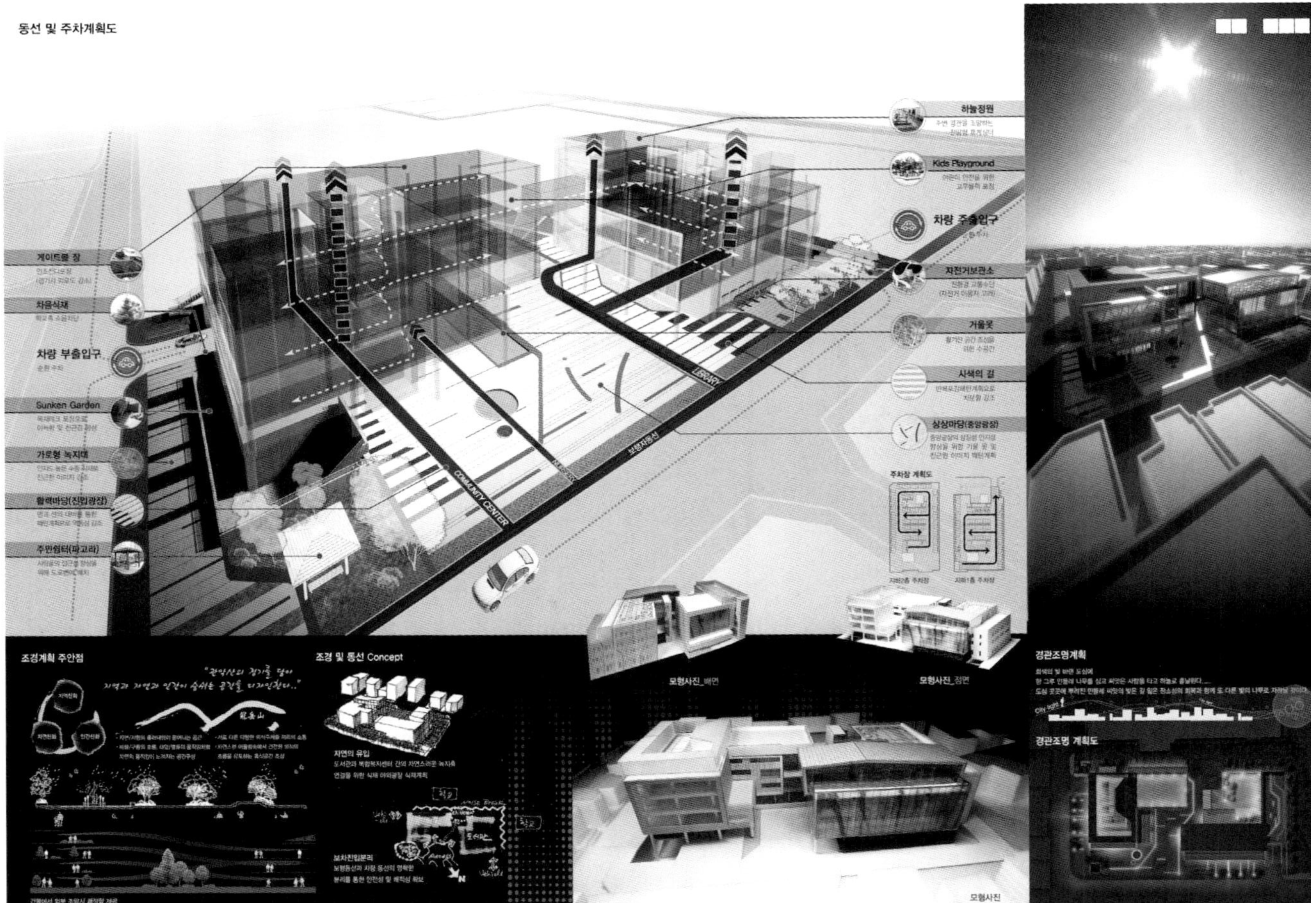

게이트볼 장
차음식재
차량 부출입구
Sunken Garden
가로형 녹지대
활력마당(진입광장)
주민쉼터(파고라)
하늘정원
Kids Playground
차량 주출입구
자전거보관소
거울못
사색의 길
심상마당(중앙광장)
주차장 계획도
지하2층 주차장
지하1층 주차장
COMMUNITY CENTER
LIBRARY
보행자동선
조경계획 주안점
조경 및 동선 Concept
자연의 유입
보차진입분리
모형사진_배면
모형사진_정면
모형사진
경관조명계획
경관조명 계획도

GWANYANG LIBRARY & WELFARE COMPLEX

Location Gwanyang 1-dong, Dongan-gu, Anyang-si, Gyeonggi-do, Korea Site Attribute General Resident Site Area 4,777.10m² Building Area 2,682.14m² Total Floor Area 15,797.16m² Landscape Area 1,870.04m² Building Coverage Ratio 56.14% Floor Area Ratio 165.07% Building Scope B2~4F Maximum Height 21.90m Structure RC Parking 236 Cars Exterior Finish Aluminium Composite Panel, T24 Pair Glass Architecture Design Kim Chi-tok | YOOSHIN Architects & Engineers, Inc. Design Team Park Sam-yong, Park Jin-ho, Park Hyeong-jun, Lee Jun-in, Kim Jong-hyeon, Park Su-a, Yeon Jeong-heum, Hong Hyeon-gi, Jeong Chung-hyeon

ATTRACTIVE LIFE - YOUR DAILY LIFE TURNS INTO AN ATTRACTIVE ONE.

Being attractive is pleasure: Pleasure gets bigger when you share it. A place where pleasure is generated and accumulated to get even bigger.
Being attractive is freedom: Three dimensional movements taking place on the rooftop and a diversity of views generated from them.
Being attractive is flexibility: The flexibility and comfortableness of a curve. An attractive and relaxing space in the city that consists of only linear lines.
Being attractive is frangrance: The pure and innocent fragrance of children, the passionate fragrance of young adults, the mild and tranquil fragrance of the aged and the fresh fragrance of nature.

LAYOUT PLAN The project was planned as an open multipurpose cultural space in consideration of the access from the local road. We brought a natural circulation for local residents by connecting the north entrance plaza and south children's playground through Banditbul playground that runs across the entire site and has been set as the path of communication linked to the residents to emphasize the function of community. There are also characteristic outer community spaces which can be accessed from Banditbul playground through an outside circulation, thus accentuating the connectivity of outer spaces and the image as a relaxation area.
FLOOR PLAN The two buildings designed for a library and welfare complex are connected through a bridge that maximizes the use of public facilities as the relaxation area and restaurant, and the zoning separated based on age increases the convenience of users. Considering the circulation of children, the lower part has a library and playhouse for children to reinforce the interconnection of children's spaces. Also, the atrium lobby and outer deck in each building draws the surrounding landscape and natural environment in the interior, harmonizing nature and the city.
ELEVATION PLAN In the city designed in a linear grid, two curved masses that symbolize nature, i.e. Gwanaksan(Mt.), are in contrast with each other to become established as landmarks of Gwanyang, Anyang City, and make a bookshelf into a pattern for the details of the elevation. Considering the daylight, the south-west side was made of insulating aluminum panel and metallic finish material to present an ultramodern image. The north-east side was entirely finished with curtain walls to avoid the direct ray of light, and the night lighting gives the light of knowledge and information to local residents.
SECTION PLAN Using the slope of the access road, the building gains a ready access, and the short mass located on the south allows a sufficient natural lighting and ventilation, bringing natural environments inside the building in an active manner. Also, a clear section zoning maintains a functional independence and the rational inner circulation pattern increases organic connectivity.

Attractive Life
일상의 삶이 매력있는 삶으로 바뀝니다.
디자인컨셉 프로세스
1ST STEP
2ND STEP
3RD STEP
"관양도서관과 복지센터는...
지역주민과 함께하는 열린 복합문화마당 입니다."
매력적인 것은 즐거움이다.
매력적인 것은 자유로움이다.
매력적인 것은 유연함이다.
매력적인 것은 향기로움이다.
GWANYANG LIBRARY
건축개요
사 업 명 관양도서관 · 복합복지센터 건립공사
위 치 안양시 동안구 관양1동 1407-1번지
지역/지구 제1종 일반주거지역
대지면적 4,777.10m²
건축면적 2,682.14m²
연 면 적 15,797.16m²
건 폐 율 56.14%
용 적 률 165.07%
구 조 철근콘크리트조
층 수 지하2층/지상4층
최고높이 21.90m
주차개요 계 : 236대(일반 177대, 확장형 45대, 장애인14대)
설비개요 EHP냉난방/공조+환기유니트 방식
외부마감 알루미늄 복합판넬, THK24 복층유리
조경개요 조경면적 : 1,870.04m²
식재면적 : 755.30m²
조경시설면적 : 1,114.74m²
북측에서 바라다본 전경

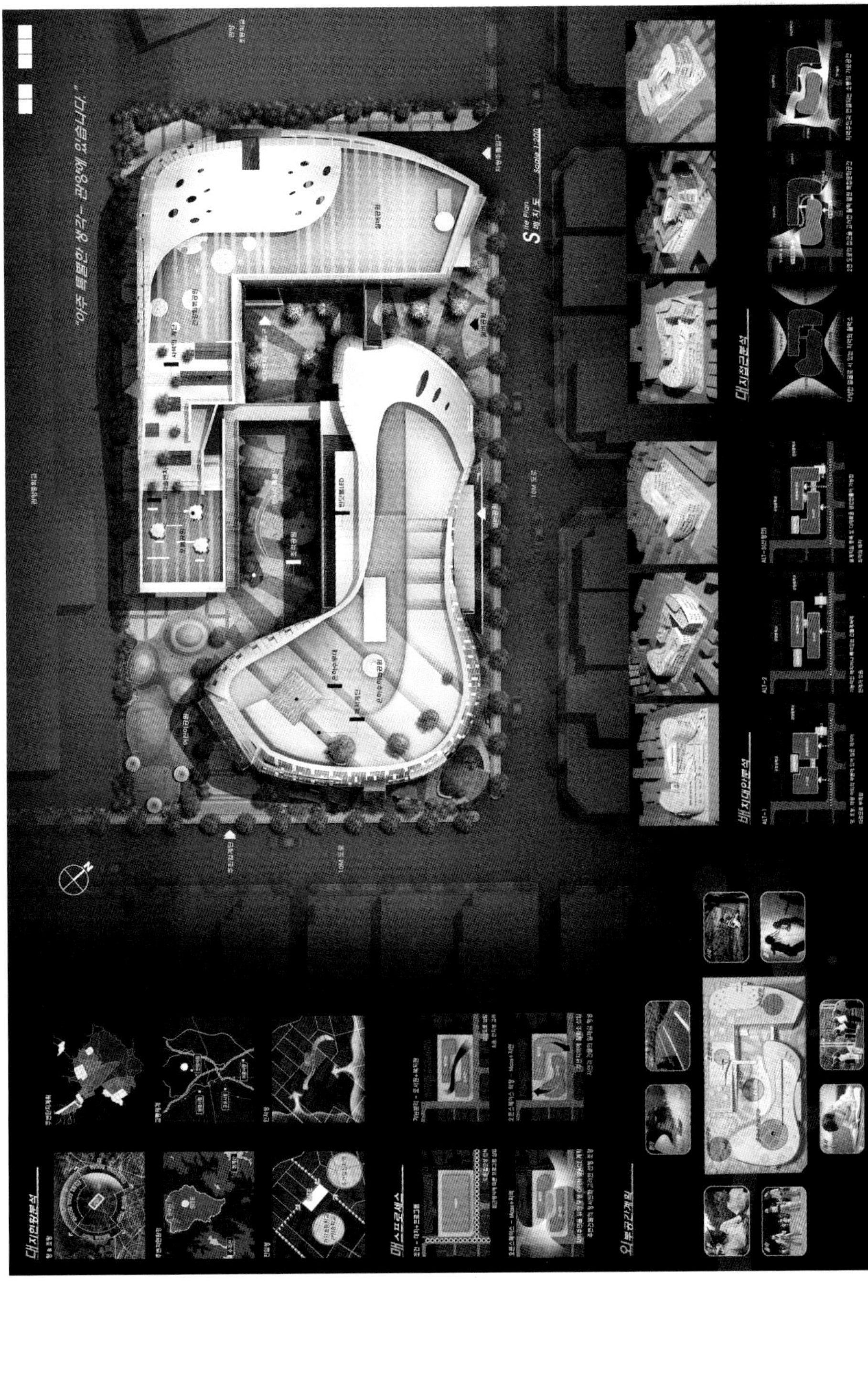

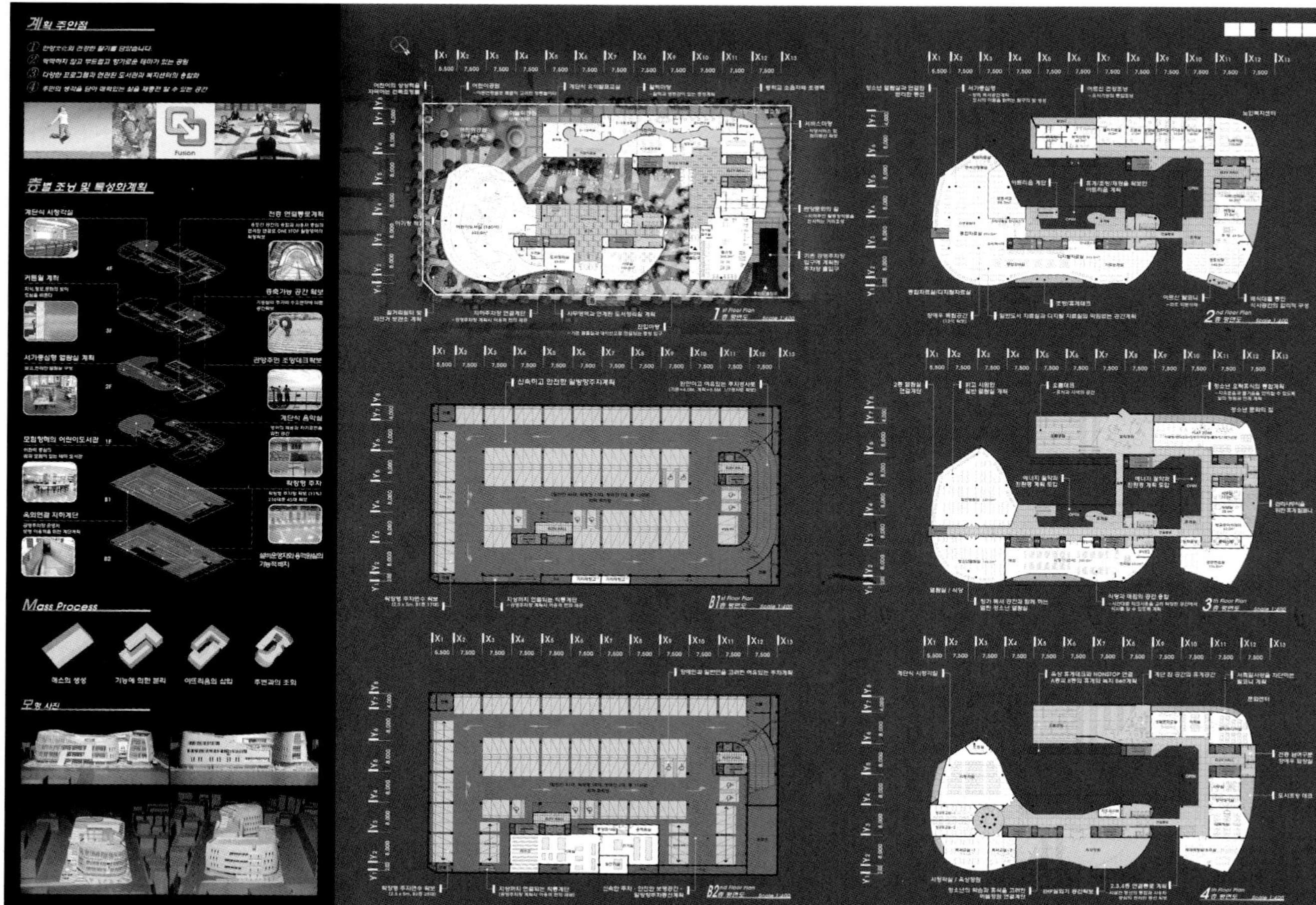
계획 주안점
Fusion
층별 조닝 및 특성화계획
Mass Process
매스의 생성
기능에 의한 분리
아트리움의 삽입
주변과의 조화
모형 사진

남측에서 바라다본 전경
입면계획
Design Motive 01 - 도서관
Design Motive 02 - 관악산
Design Motive 03 - 한양천
스킨계획
X13 X12 X11 X10 X9 X8 X7 X6 X5 X4 X3 X2 X1
7,500 7,500 7,500 7,500 7,500 7,500 7,500 7,500 7,500 7,500 7,500 5,500
Y6 Y5 Y4 Y3 Y2 Y1
8,000 8,000 8,000 8,000 8,000
T4 알루미늄 복합판넬(비오암 시스템)
T24 칼라복층유리
South Elevation 남측 입면도 Scale 1:400
East Elevation 동측 입면도 Scale 1:400
X1 X2 X3 X4 X5 X6 X7 X8 X9 X10 X11 X12
5,500 7,500 7,500 7,500 7,500 7,500 7,500 7,500 7,500 7,500 7,500 7,500
Y1 Y2 Y3 Y4 Y5 Y6
8,000 8,000 8,000 8,000 8,000 4,000
North Elevation 북측 입면도 Scale 1:400
West Elevation 서측 입면도 Scale 1:400

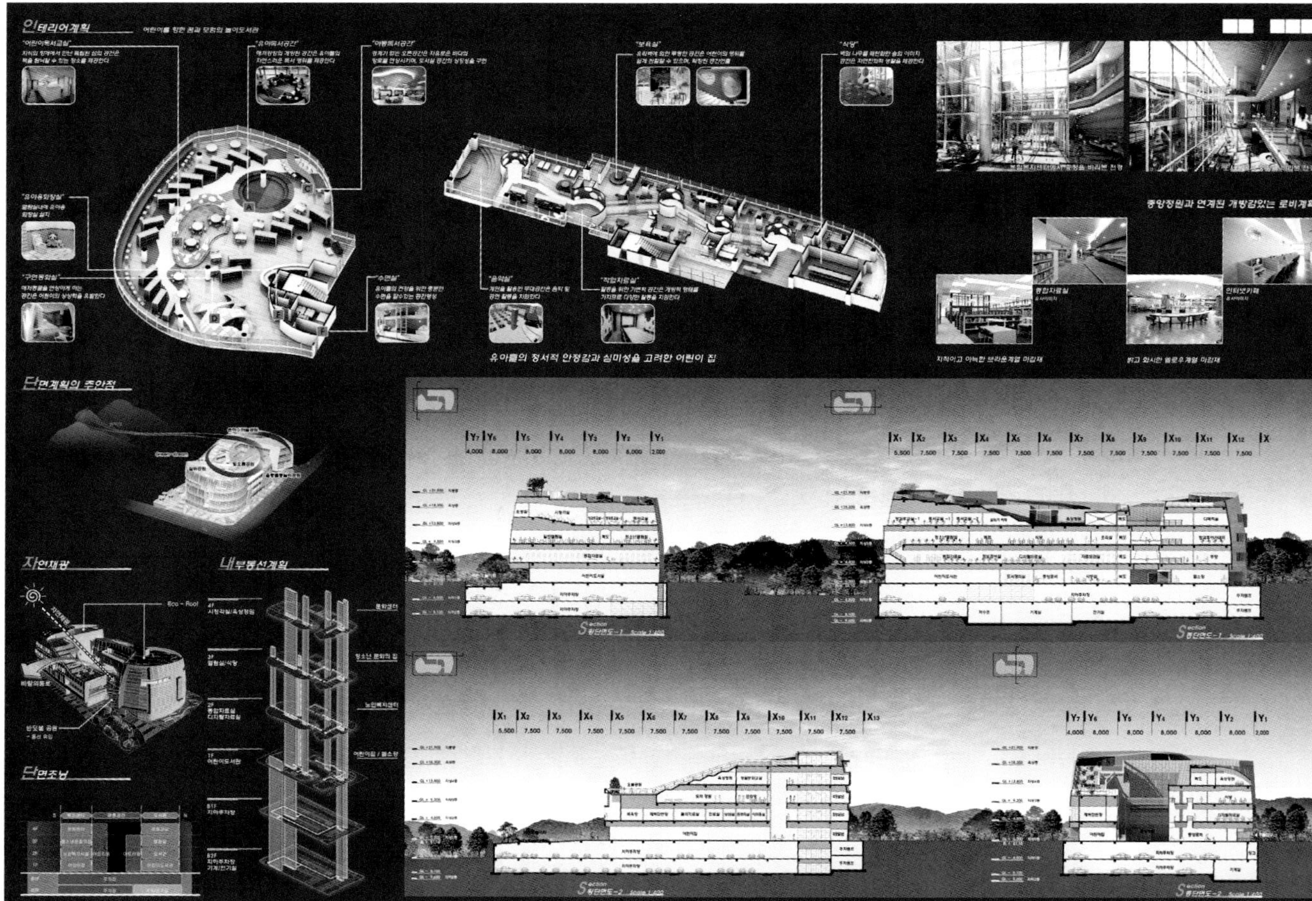
인테리어계획
유아들의 정서적 안정감과 심미성을 고려한 어린이 집
중앙정원과 연계된 개방감있는 로비계획
단면계획의 주안점
자연채광
내부동선계획
단면조닝
Section 횡단면도-1
Section 종단면도-1
Section 횡단면도-2
Section 종단면도-2

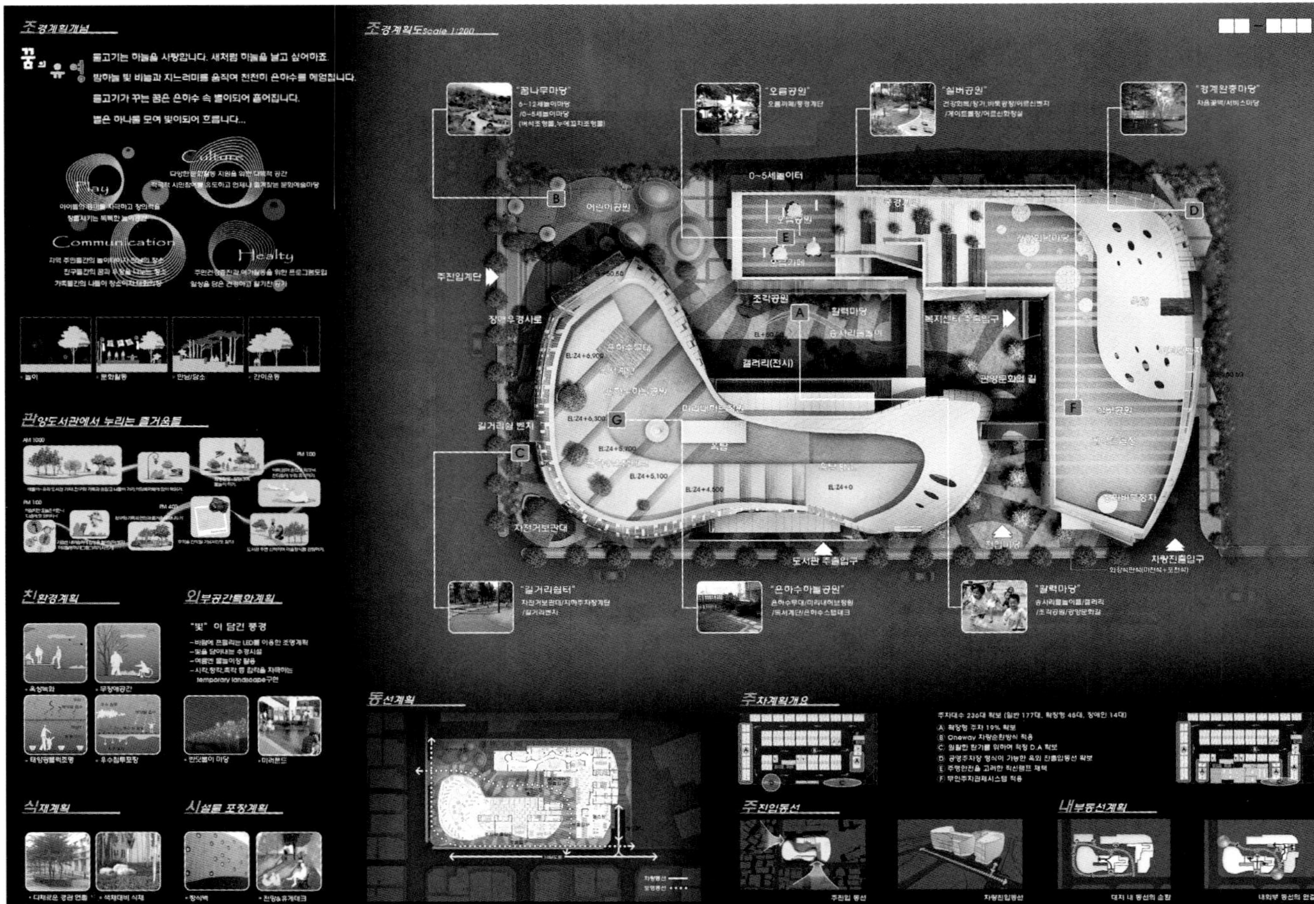
조경계획개념
꿈의 유영
물고기는 하늘을 사랑합니다. 새처럼 하늘을 날고 싶어하죠.
밤하늘 및 비늘과 지느러미를 움직여 천천히 은하수를 헤엄칩니다.
물고기가 꾸는 꿈은 은하수 속 별이되어 흩어집니다.
별은 하나둘 모여 빛이되어 흐릅니다...
Culture
Play
Communication
Healty
관양도서관에서 누리는 즐거움들
친환경계획
외부공간특화계획
"빛" 이 담긴 풍경
식재계획
시설물 포장계획
조경계획도 Scale 1:200
"꿈나무마당"
"오름공원"
"실버공원"
"경계완충마당"
0~5세놀이터
어린이공원
주진입계단
조각공원
활력마당
갤러리(전시)
복지센터 주출입구
관양문화의 길
길거리쉼 벤치
자전거보관대
도서관 주출입구
차량진출입구
"길거리쉼터"
"은하수하늘공원"
"활력마당"
동선계획
주차계획개요
주진입동선
내부동선계획

THE LIBRARY OF CHANGWON NATIONAL UNIVERSITY

Location Changwon-si, Gyeongsangnam-do, Korea **Site Attribute** Natural Green, School Facilities Decision **Site Area** 13,088m² **Building Area** 4,716.6m² **Total Floor Area** 15,688.6m² **Building Coverage Ratio** 36.0% **Floor Area Ratio** 119.9% **Building Scope** B1~5F **Maximum Height** 24.1m **Structure** RC, Steel **Parking** 63 Cars **Exterior Finish** T24 Color Pair Glass, Ceramics Panel, Ceramics Louver **Architecture Design** Eun Dong-sin, Hong Keun-pyo | EGA Architecture Co., Ltd. + Kim Chun-gu | TODAM Architects **Design Team** Yu Myeong-sun, Seok Min-cheol, Kim Yeong-mon, Kim Su-jin

PROLOGUE In the period which a digital informatization is progressed, to contain a shape of the existing library is only a museum of books.

A book is past, but we see the future through it.

The library is exchanged for the information by an action but not only an existence of the space and, in various movements and the corrosion of an outer condition to change randomly, is finished as a movement of an infinite changing information activity. It is a place to be a center to offer a new type of information service in the new time and a boundary space to exchange and mediate the inside and the outside.

SITE The campus of the Changwon University located at the north-east in Changwon city has a backyard surrounded by the Bonglim mountain and looks down the downtown of the Changwon in the front. As a plan site is connected with main roadside to link the main entrance and the back gate (a dormitory), it is required for a consideration of a recognition and an access in them. Also, it is placed at the boundary which a linear horizontal system of the campus and a natural factor in the garden near the plan site collide.

MASS The library of Changwon University forms a Mass by the nearby environmental factors by the collision of the linear system of the campus and the free curve of the nature. The natural flow curve of the garden penetrates and the central garden is formed, and the in the front, an extended flow of the land gives a space with the frontness and the recognition. By a collision of such different forces, a form of the Mass changes.

COMMUNICATION As, in the central garden formed by the nature's curve, a surface area to meet the outside from the inside of the library gets wider, a communication space to give a visual openness and to be with a nature is enlarged. The central garden is an outer space and forms a central sense of space, an inner space, and the sphere of the inner library is expanded. The large space formed inside divides the spheres according to a function and unifies the spheres through a cross of the eyes. This cross of eyes symbolizes a share of information and the large space becomes a place to exchange information.

LIBRARY AS THE LOCAL CENTRAL FACILITY In this plan, a library is a local central space and required for a role as an open space for the students as well as the citizens of Changwon. To accept such a demand, the space is divided in the open sphere and the BDS (data loss prevention system) sphere. The open spheres including a reading room, an exhibition room, a lobby and an internet cafe makes it possible for the students and the citizens to use freely and acquires a publicity. There is no direct connection for the information exchange through the media of books. While a lot of media including a book, a man, a space and a nature are crossed through a space, the information sometimes flows and sometimes are contained through the mutual communication. I hope that such a free flow of information and a sea of information become an appearance of the library.

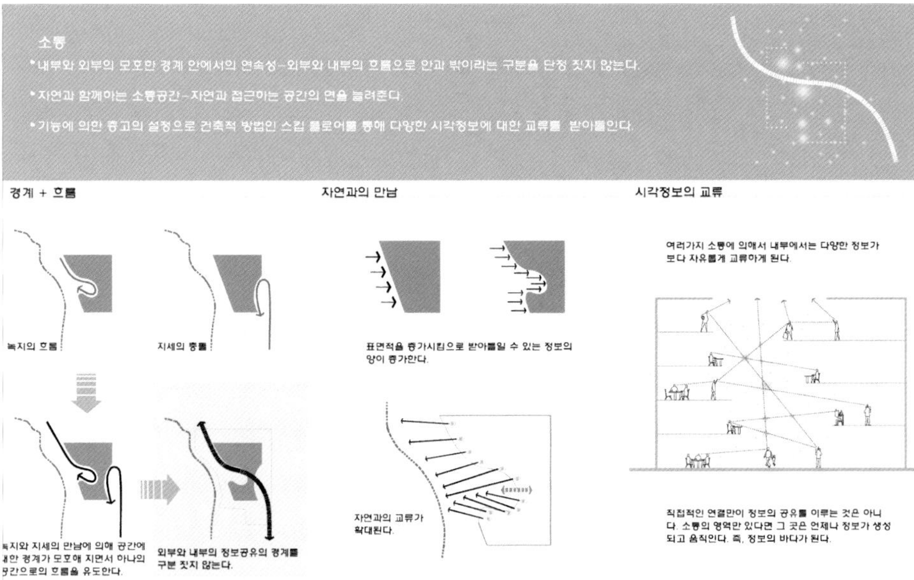

SPACE CONCEPT

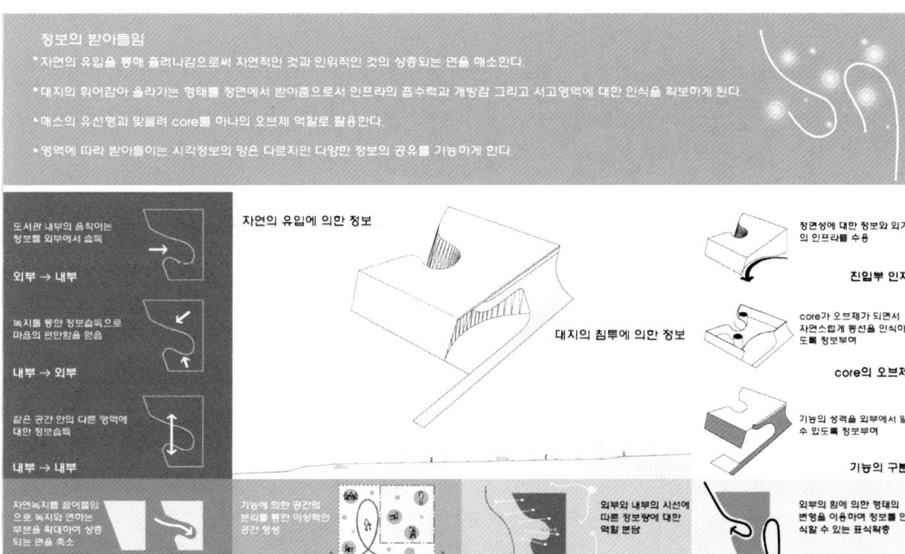

DESIGN PROCESS

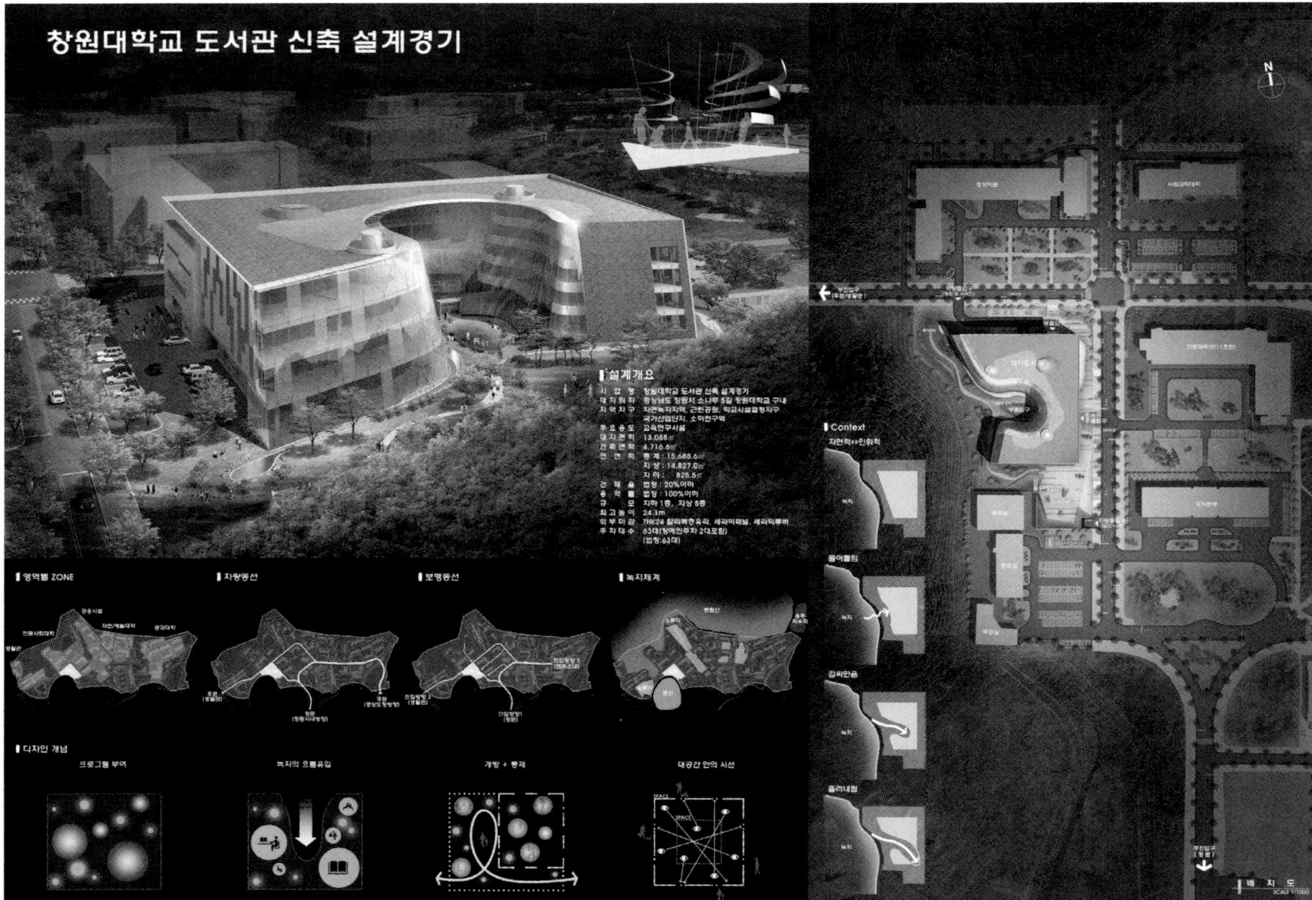
창원대학교 도서관 신축 설계경기
설계개요
사 업 명 창원대학교 도서관 신축 설계경기
대 지 위 치 경상남도 창원시 소나무 5길 창원대학교 구내
지 역 지 구 자연녹지지역, 근린공원, 학교시설결정지구
국가산업단지, 소하천구역
주 요 용 도 교육연구시설
대 지 면 적 13,088㎡
건 축 면 적 4,716.6㎡
연 면 적 총 계 : 15,688.6㎡
지 상 : 14,827.0㎡
지 하 : 825.5㎡
건 폐 율 법정 : 20%이하
용 적 률 법정 : 100%이하
규 모 지하 1층, 지상 5층
최 고 높 이 24.1m
외 부 마 감 THK24 칼라복층유리, 세라믹패널, 세라믹루버
주 차 대 수 63대(장애인주차 2대포함)
(법정:63대)
영역별 ZONE
차량동선
보행동선
녹지체계
디자인 개념
프로그램 부여
녹지의 흐름유입
개방 + 통제
대공간 안의 시선
SPACE
Context
자연적↔인위적
녹지
배 치 도
SCALE 1/1000

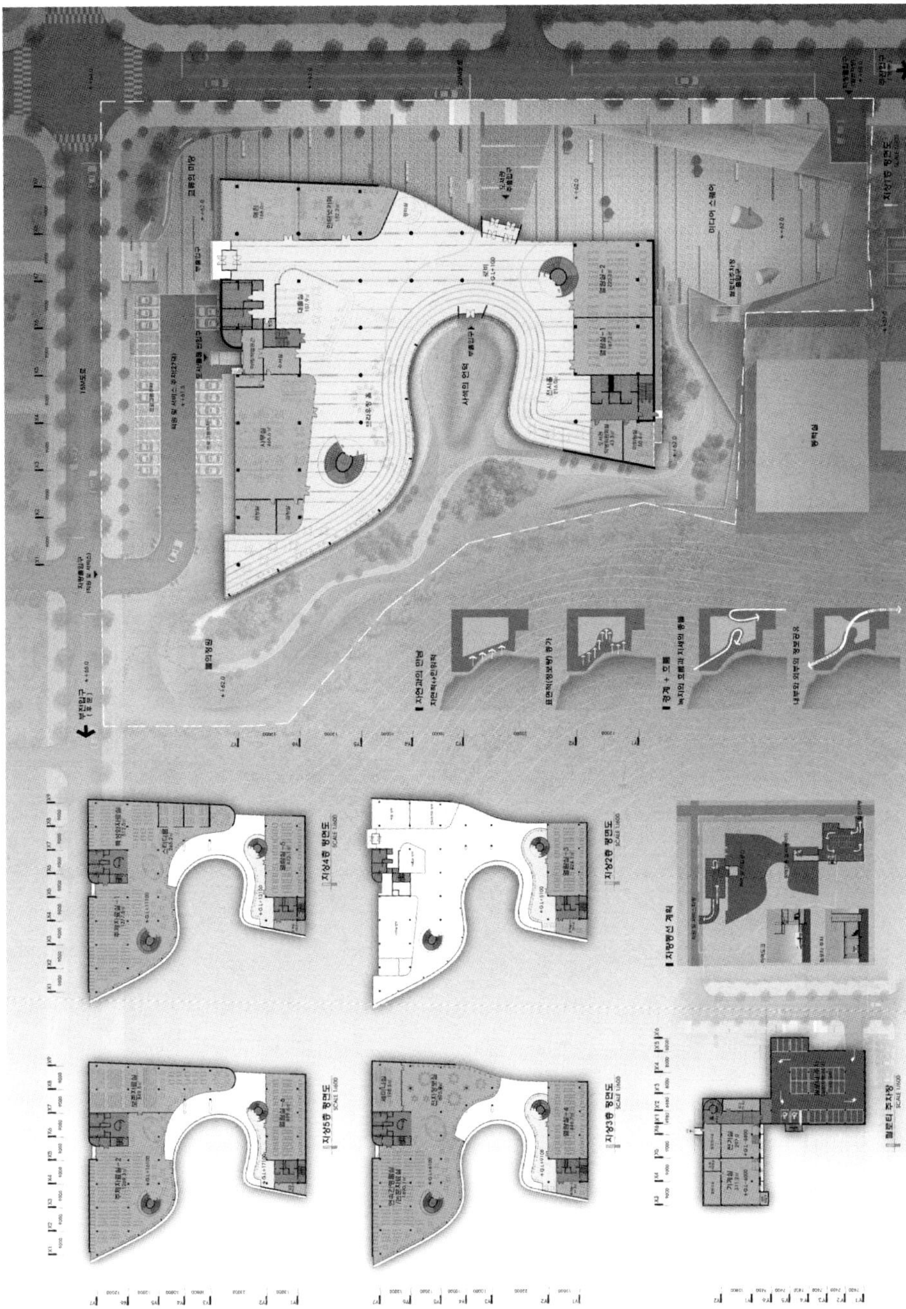

정 면 도
SCALE 1/400
우 측 면 도
SCALE 1/400
배 면 도
SCALE 1/400
좌 측 면 도
SCALE 1/400
횡 단 면 도
SCALE 1/400
단면개념
종 단 면 도
SCALE 1/300

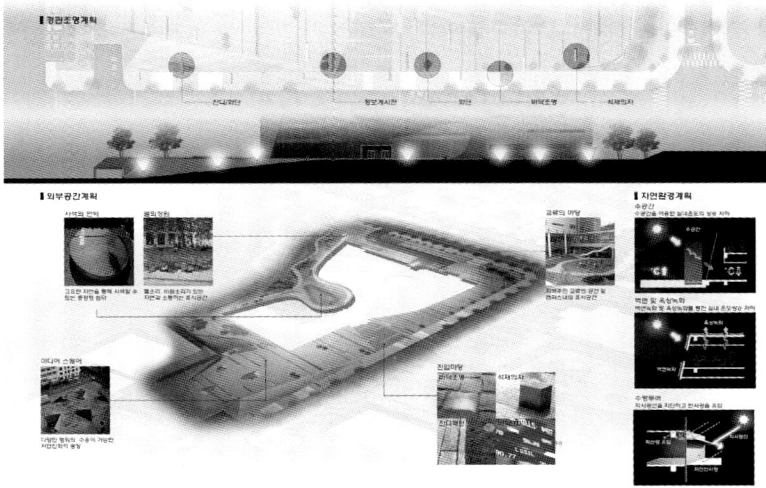

디지털,정보화가 진행되는 시대에 기존 도서관의 모습만을 담는다면
그것은 책박물관에 불과하다.

"책은 과거이지만 우리는 책을 통해 미래를 본다."

도서관은 미래를 향한 새로운 공간이어야 한다.

>> EDUCATION

THE LIBRARY OF CHANGWON NATIONAL UNIVERSITY

Location Changwon-si, Gyeongsangnam-do, Korea **Site Attribute** Natural Green **Site Area** 13,088m² **Building Area** 5,303m² **Total Floor Area** 15,821m² **Building Scope** B1~6F **Maximum Height** 28.5m **Structure** SRC, Steel **Parking** 64 Cars **Architecture Design** Cho Ju-hwan | SIAPLAN Architects & Planners **Design Team** Yun Jeong-hyeon, Yang Hui-beom, Seo Seong-deok, Kim Du-yeol, Choi Jeong-su, Choi Bom-i

The library of Changwon National University to be built so as to carry out a key role of the knowledge information, culture, leisure and lifelong study education of the local society is the landmark of the campus, exceeds its function of reading the books and will be reborn as a place for a university culture contents.

CONCEPT OF PLAN

Open Library (Openness)

- Plan for a place of knowledge culture to provide the study and the cutting edge information as a library with new concept through forming an open big space
- Landmark: Plan for a symbolic building within the campus without turning off a light for 24 hours
- Ecology: Plan for a green building possible for a comfortable indoor environment and an energy saving by inducing the neighboring greens inside.

DESIGN PROCESS

- Floating: Constituting an open mass by inducing the greens of the west in the campus and accepting a walking traffic line to cross the site
- Array: Re-arraying the mass connected to the greens by setting up a symbolic nature's axis to link the Najeongbyeong peak and the Jangbok mountain.
- Opening: Plan for an open space to open the visual axis of the green and to connect the outer space of the campus
- Adaptation: Plan for the community space to accept the lattice-type road axis and consider the campus context
- Symbol: By planning a symbolic atrium to recognize in entering the main gate, realizing an image of the light of knowledge in the campus not to turn off for 24 hours
- Intrusion: Plan for the green carpet to link the neighboring greens flow and plan for the eco tube to elevate the greens vertically

BLOCK PLANNING The library located between the Human Science building and the Economic & Business building within Changwon National University acquires the visual axis of the greens of the campus to the maximum, adapts to the context of the nature and the building and considers a new construction and an extension in the future. In addition, the atrium of the landmark considering a recognition and a symbolism in entering the main gate is planned and through a large space of the atrium, an image of open library spatially and visually is realized.

PLANNING The core only for the reading room is planned so that it is possible to open for 24 hours through zoning of each floor of the reading room and the reference room and by the entry of each floor into the first and second floors, the spaces with various entering parts are constituted and the traffic line for the users and the traffic line for the service are divided and it is considered for an access and a convenience. In addition, by planning a modularized bookshelf, it is planned to apply reasonably the library's inherent program.

ELEVATION/SECTION PLAN The horizontal and vertical louver considering a direction and a irradiance and the elevation design to emphasize a transparency are to emphasize a visual connection of the inner and outer spaces, and by concentrating the resting facility, media facility and greens on the central core and making the green core, it is planned for a comfortable indoor environment. In addition, through atrium, it is considered for a vertical space connection and a visual openness.

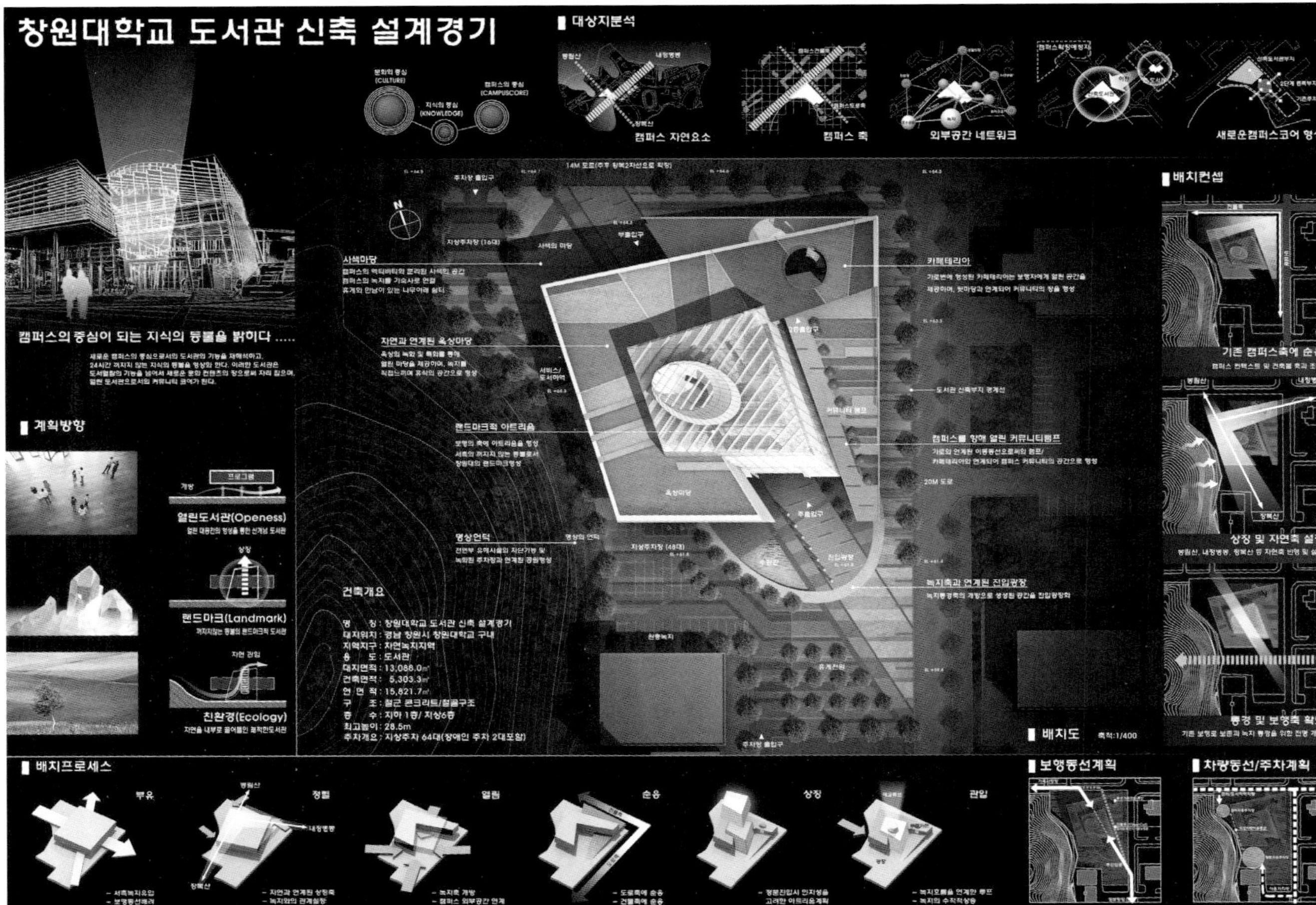
창원대학교 도서관 신축 설계경기
문화의 중심 (CULTURE)
지식의 중심 (KNOWLEDGE)
캠퍼스의 중심 (CAMPUSCORE)
대상지분석
캠퍼스 자연요소
캠퍼스 축
외부공간 네트워크
새로운캠퍼스코어 형성
캠퍼스의 중심이 되는 지식의 등불을 밝히다 .....
새로운 캠퍼스의 중심으로서의 도서관의 기능을 재해석하고,
24시간 꺼지지 않는 지식의 등불을 형상화 한다. 이러한 도서관은
도서열람의 기능을 넘어서 새로운 문화 컨텐츠의 장으로써 자리 잡으며,
열린 도서관으로서의 커뮤니티 코어가 된다.
계획방향
열린도서관(Openess)
랜드마크(Landmark)
친환경(Ecology)
14M 도로(추후 왕복2차선으로 확장)
주차장 출입구
지상주차장 (16대)
사색의 마당
사색마당
캠퍼스의 액티비티와 분리된 사색의 공간
캠퍼스의 녹지를 가속사로 연결
휴게와 만남이 있는 나무아래 쉼터
자연과 연계된 옥상마당
옥상의 녹화 및 휘화를 통해
열린 마당을 제공하며, 녹지를
직접느끼며 휴식의 공간으로 형성
서비스/
도서하역
랜드마크적 아트리움
보행의 축에 아트리움을 형성
서측의 꺼지지 않는 등불로서
창원대의 랜드마크형성
영상언덕
전면부 유해시설의 차단기능 및
녹화된 주차장과 연계된 공원형성
건축개요
명 칭 : 창원대학교 도서관 신축 설계경기
대지위치 : 경남 창원시 창원대학교 구내
지역지구 : 자연녹지지역
용 도 : 도서관
대지면적 : 13,088.0㎡
건축면적 : 5,303.3㎡
연 면 적 : 15,821.7㎡
구 조 : 철근 콘크리트/철골구조
층 수 : 지하 1층/ 지상6층
최고높이 : 28.5m
주차개요 : 지상주차 64대(장애인 주차 2대포함)
카페테리아
가로변에 형성된 카페테리아는 보행자에게 열린 공간을
제공하며, 뒷마당과 연계되어 커뮤니티의 장을 형성
도서관 신축부지 경계선
캠퍼스를 향해 열린 커뮤니티램프
가로와 연계된 이동동선으로써의 램프/
카페테리아와 연계되어 캠퍼스 커뮤니티의 공간으로 형성
20M 도로
녹지축과 연계된 진입광장
녹지공간의 개방으로 생성된 공간을 진입광장화
옥상마당
지상주차장 (48대)
주차장 출입구
배치도 축척:1/400
배치컨셉
기존 캠퍼스축에 순응
상징 및 자연축 설정
통경 및 보행축 확보
배치프로세스
부유
정렬
열림
순응
상징
관입
보행동선계획
차량동선/주차계획

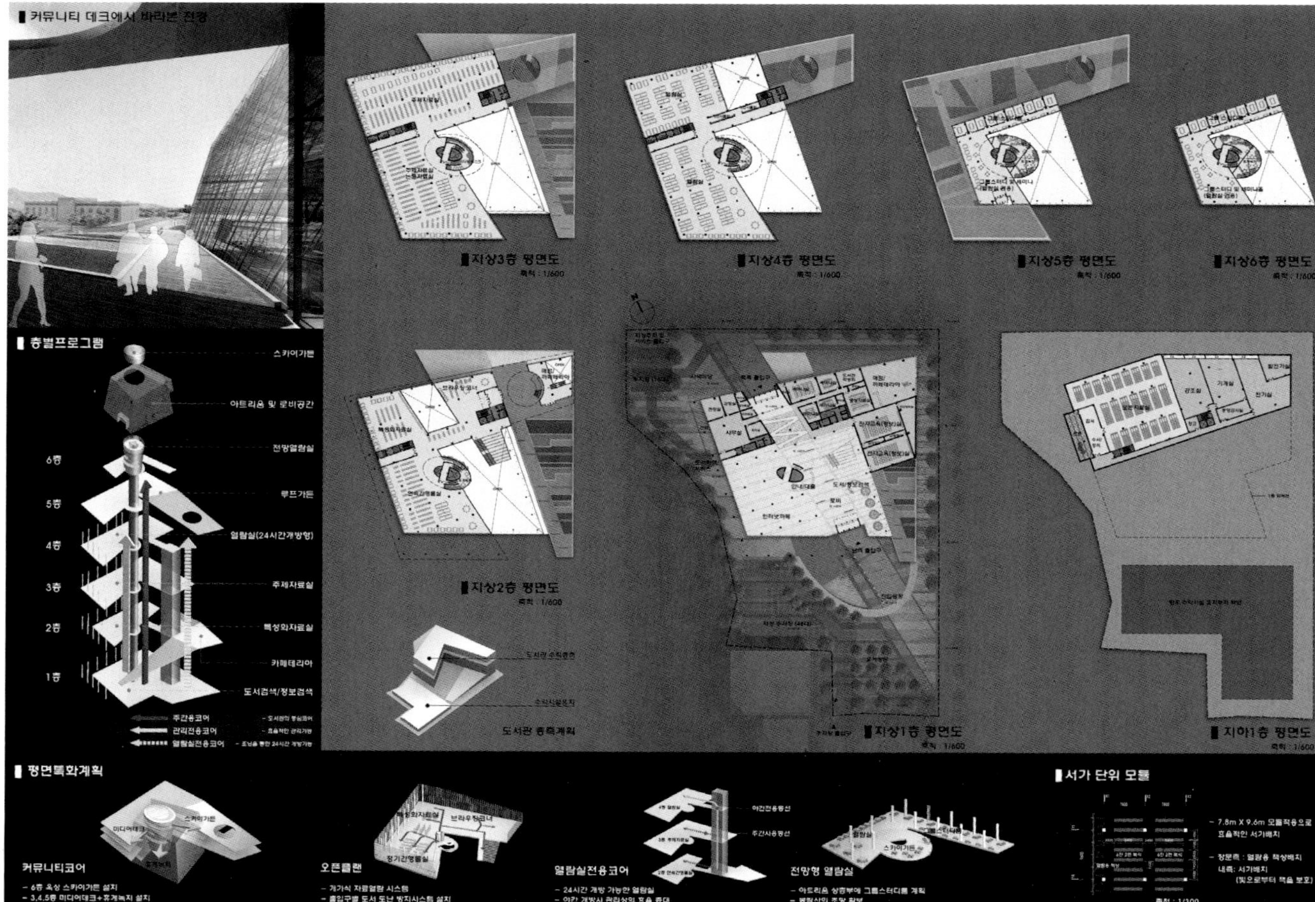
커뮤니티 데크에서 바라본 전경
지상3층 평면도
축척 : 1/600
지상4층 평면도
축척 : 1/600
지상5층 평면도
축척 : 1/600
지상6층 평면도
축척 : 1/600
층별프로그램
스카이가든
아트리움 및 로비공간
전망열람실
루프가든
열람실(24시간개방형)
주제자료실
특성화자료실
카페테리아
도서검색/정보검색
6층
5층
4층
3층
2층
1층
주간용코어
관리전용코어
열람실전용코어
지상2층 평면도
축척 : 1/600
도서관 층축계획
지상1층 평면도
축척 : 1/600
지하1층 평면도
축척 : 1/600
평면특화계획
커뮤니티코어
- 6층 옥상 스카이가든 설치
- 3,4,5층 미디어데크+휴게복지 설치
오픈플랜
- 개가식 자료열람 시스템
- 출입구별 도서 도난 방지시스템 설치
열람실전용코어
- 24시간 개방 가능한 열람실
- 야간 개방시 관리상의 효율 증대
전망형 열람실
- 아트리움 상층부에 그룹스터디룸 계획
- 봉림산의 조망 확보
서가 단위 모듈
- 7.8m X 9.6m 모듈적용으로 효율적인 서가배치
- 창문측 : 열람용 책상배치
내측 : 서가배치
(빛으로부터 책을 보호)
축척 : 1/300

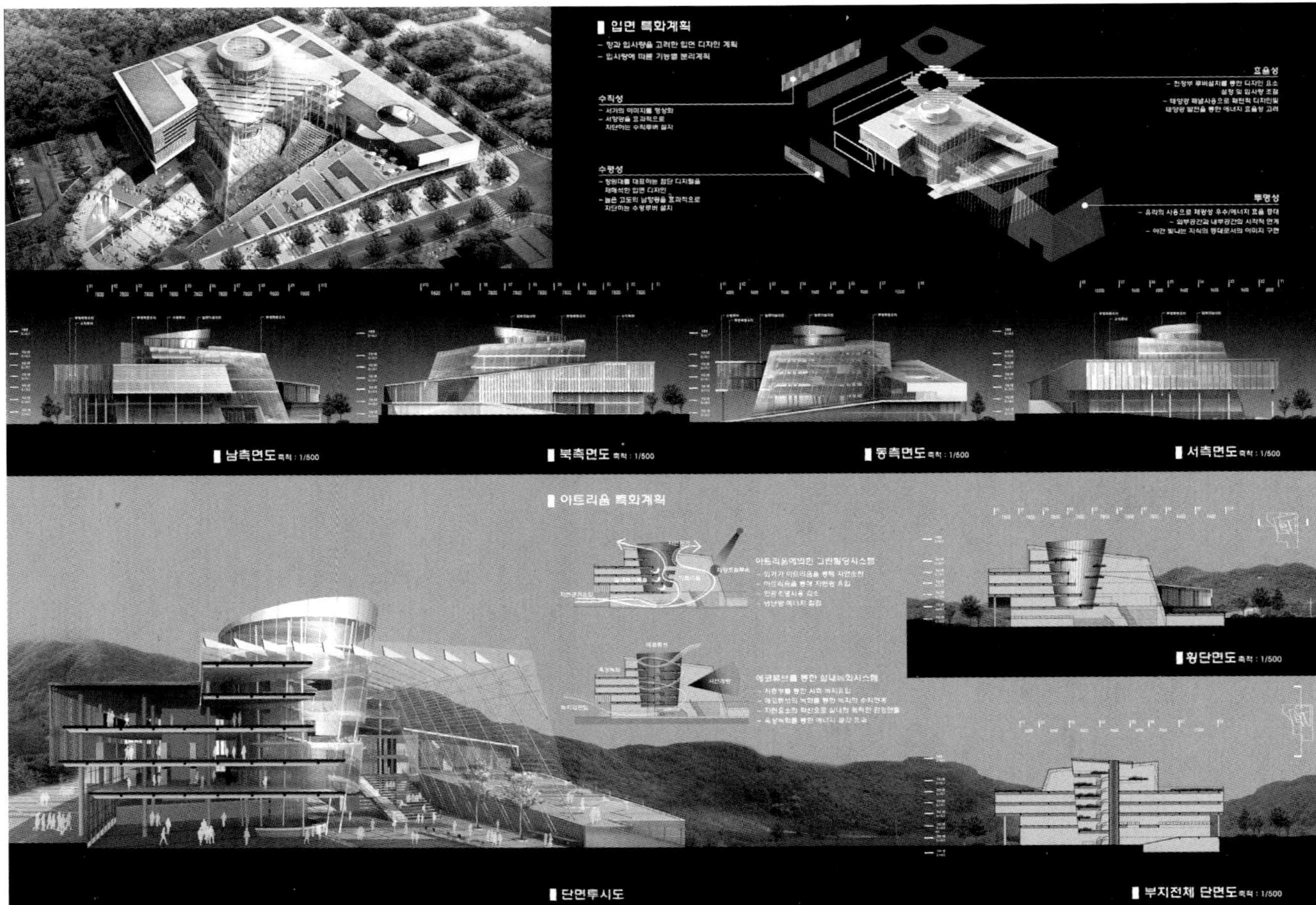
입면 특화계획
– 향과 입사량을 고려한 입면 디자인 계획
– 입사량에 따른 기능별 분리계획
수직성
– 서가의 이미지를 형상화
– 서양광을 효과적으로 차단하는 수직루버 설치
수평성
– 창원대를 대표하는 첨단 디지털을 재해석한 입면 디자인
– 높은 고도의 남향광을 효과적으로 차단하는 수평루버 설치
효율성
– 천정부 루버설치를 통한 디자인 요소 설정 및 입사량 조절
– 태양광 패널사용으로 패턴적 디자인및 태양광 발전을 통한 에너지 효율성 고려
투명성
– 유리의 사용으로 채광성 우수/에너지 효율 증대
– 외부공간과 내부공간의 시각적 연계
– 야간 빛나는 지식의 등대로서의 이미지 구현
남측면도 축척 : 1/500
북측면도 축척 : 1/500
동측면도 축척 : 1/500
서측면도 축척 : 1/500
아트리움 특화계획
아트리움에의한 그린빌딩시스템
에코루브를 통한 실내녹화시스템
횡단면도 축척 : 1/500
단면투시도
부지전체 단면도 축척 : 1/500

SEONGNAM PANGYO ECOLOGY EXHIBITION HALL

Location Pangyo-dong, Bundang-gu, Seongnam-si, Gyeonggi-do, Korea **Site Attribute** Natural Green **Site Area** 2,149.00m² (Neighborhood Park Area 95,968m²) **Building Area** 1,881.48m² **Total Floor Area** 3,999.21m²(Ground 2,498.15m², Underground 1,501.06m²) **Building Coverage Ratio** 1.96% **Floor Area Ratio** 2.60% **Building Scope** B1~2F **Maximum Height** 17.1m **Structure** RC **Main Finish** Low-E Color Pair Glass, Exposed Concrete, AL Punching Plate, Wood, Natural Stone Net **Architecture Design** Jeong Young-kyoon, Park Gyu-yong | HEERIM Architects & Planners **Design Team** Song Gwang-uk, Choi Sun-ju, Kim Ki-beom, Oh Seon-hwa, Kang Min-ju

Under the design concept of 'a box to hold nature', the project features efficient spatial composition of an exhibition area, education / information area and a greenhouse that are managed independently yet are connected when in need.

THE BASIC DIRECTION OF THE PLAN AIMS AT First, a field study area to experience ecosystem. We designed a space where nature study from the park is extended to give direct experience in an ecological hall (greenhouse) and exhibition area, combining the education and exhibition. Second, an environment-friendly space. The ecological hall (greenhouse) is located inside the building to function as a pleasant green courtyard, and the nature-friendly materials of its elevation and the sky garden on the rooftop create an environment-friendly space. Third, a space of local symbol. We focused on presenting an ecological space as a part of nature, a symbolic space of independent identity linked to the park. Fourth, a joyful event space. The design follows the intention to provide an open space where enjoyable events can take place as well as exhibition and education programs.

EXHIBITION SPACE CONCEPT

"The environment created by nature and human"

Spatial approach symbolizing a diagram composed of lines and planes based on the basic element of a life language that brings balances to nature.

· The Window of Environment - Urban Environment: The program, which concerns the influence of human behavior on nature, is presented by a communicative language shared by the ecosystem and human, establishing a linear symmetrical relation.

· The Window of Ecology - Ecological Environment: The Circles symbolizing the movement, expansion and continuous response in nature visualize the life of ecological pulse by representing the spirit and message of ecosystem with various means of intervention and interaction.

· The Window of Life - Hydro Environment: The object derived from water, which is an element of life, is represented three-dimensionally in a modern space as an independent ecological component in the flow of environment that changes in an organic manner.

다 함께, 에코워크! _ 참여하고 바꾸며 실천하는 삶 배우기
멀티 디오라마 영상
미디어 테이블

환경의 창 _ 환경기초시설과 연계된 환경과 생활의 배움터

생태의 창 _ 성남의 동물과 식물에 대한 종합적 정보와 생태 배우기

자연을 담는 상자
성남판교 생태학습원 건축설계
대지 SITE
배치 개념 SITE PLAN CONCEPT
주변현황 CONTEXT
개요 SUMMARY
《주 진입
《차량 진입
하늘정원 진입》
형태 구축 DESIGN PROCESS
base cube
cube + eco
태양각도에 따른 지붕라인 형성
cube for nature
친환경 계획 SUSTAINABLE DESIGN

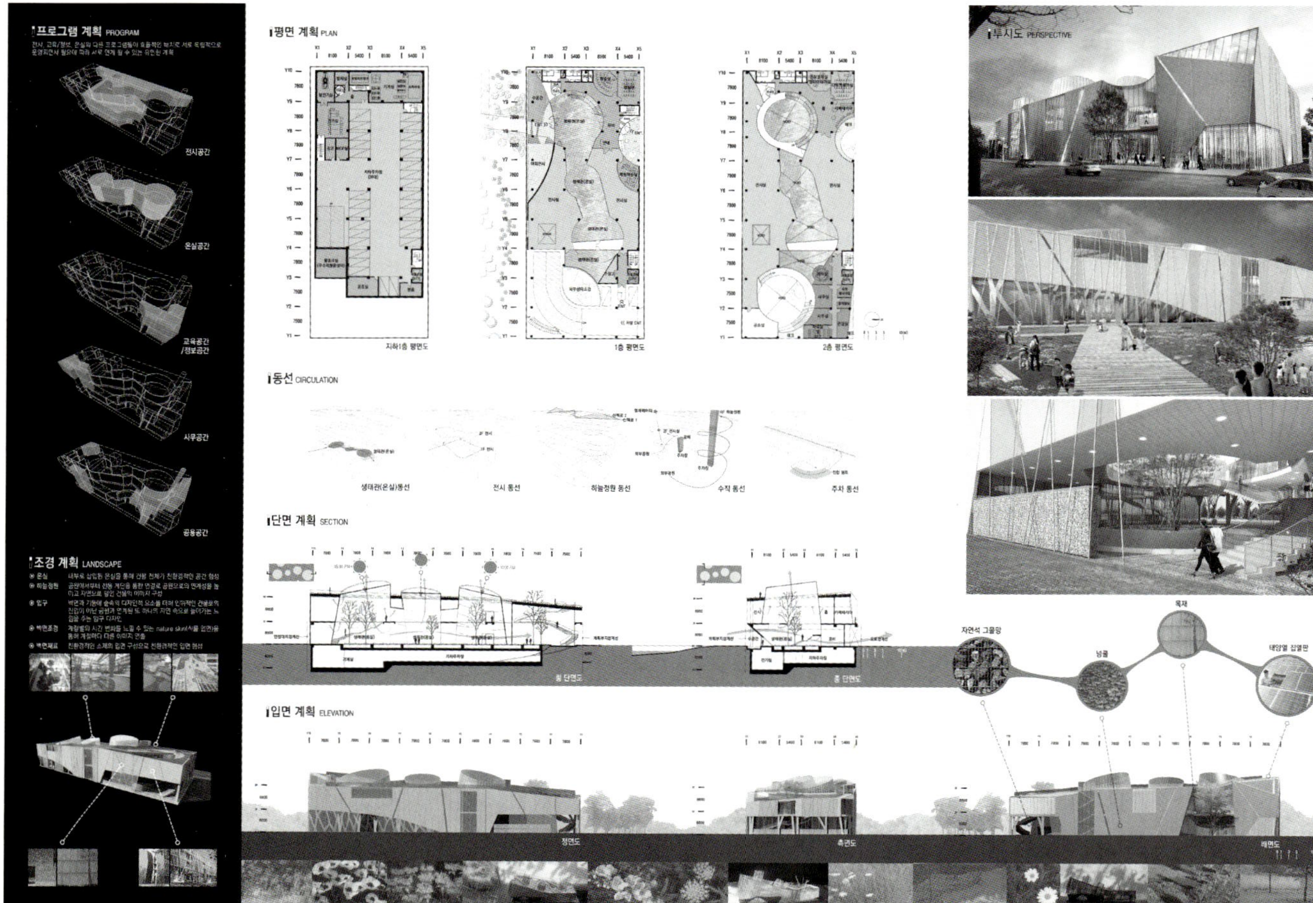
프로그램 계획 PROGRAM
전시, 교육/정보, 온실과 다른 프로그램들이 효율적인 배치로 서로 독립적으로 운영되면서 필요에 따라 서로 연계 될 수 있는 유연한 계획
전시공간
온실공간
교육공간 /정보공간
사무공간
공용공간
조경 계획 LANDSCAPE
온실 내부로 삽입된 온실을 통해 건물 전체가 친환경적인 공간 형성
하늘정원 공원에서부터 진입 계단을 통한 연결로 공원으로의 연계성을 높이고 자연으로 덮인 건물의 이미지 구성
입구 벽면과 기둥에 숲속의 디자인적 요소를 더해 인위적인 건물로의 진입이 아닌 공원과 연계된 또 하나의 자연 속으로 들어가는 느낌을 주는 입구 디자인
벽면조경 계절별의 시간 변화를 느낄 수 있는 nature skin식을 입면을 통해 계절마다 다른 이미지 연출
벽면재료 친환경적인 소재의 입면 구성으로 친환경적인 입면 형성
평면 계획 PLAN
지하1층 평면도
1층 평면도
2층 평면도
동선 CIRCULATION
생태관(온실)동선
전시 동선
하늘정원 동선
수직 동선
주차 동선
단면 계획 SECTION
횡 단면도
종 단면도
입면 계획 ELEVATION
정면도
측면도
배면도
투시도 PERSPECTIVE
자연석 그물망
낭굴
목재
태양열 집열판